Collins

KS3 Science

Book 2

Tyreece + Tyra Parkes

D0320331

Ed Walsh
Series Editor
Tim Greenway
Ray Oliver
David Taylor

William Collins' dream of knowledge for all began with the publication of his first book in 1819. A self-educated mill worker, he not only enriched millions of lives, but also founded a flourishing publishing house. Today, staying true to this spirit, Collins books are packed with inspiration, innovation and a practical expertise. They place you at the centre of a world of possibility and give you exactly what you need to explore it.

Collins. Freedom to teach.

Published by Collins
An imprint of HarperCollinsPublishers
The News Building
1 London Bridge Street
London
SE1 9GF

Browse the complete Collins catalogue at
www.collinseducation.com

© HarperCollinsPublishers Limited 2008

12

ISBN-13 978-0-00-726 421-6

ISBN-10 0-00-726421-6

All rights reserved. No part of this publication may be reproduced, stored in a retrieval system or transmitted in any form or by any means - electronic, mechanical, photocopying, recording or otherwise - without the prior written consent of the Publisher or a licence permitting restricted copying in the United Kingdom issued by the Copyright Licensing Agency Ltd, 90 Tottenham Court Road, London W1T 4LP.

British Library Cataloguing in Publication Data. A Catalogue record for this publication is available from the British Library.

Commissioned by Cassandra Birmingham
Project managed by Penny Fowler and Anne Summers
Glossary written by Pam Large
Edited and proofread by Camilla Behrens, Rachel Hutchings,
Lynn Watkins, Rosie Parrish and Anita Clark
Internal design by Jordan Publishing Design
Page layout by eMC Design Ltd, www.emcdesign.org.uk
Illustrations by Jorge Santillan, Peters & Zabransky

Production by Arjen Jansen
Printed and bound by CPI Group (UK) Ltd, Croydon, CR0 4YY

Contents

Introduction

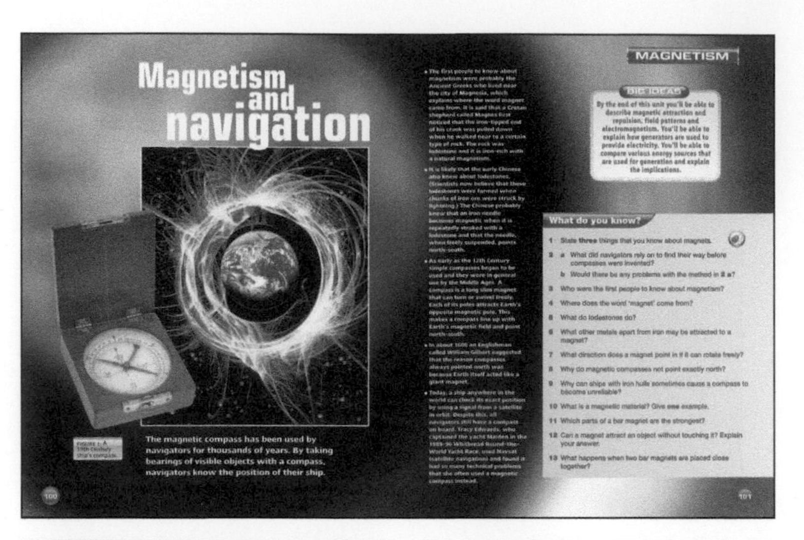

Exciting Topic Openers

Every topic begins with a fascinating and engaging article to introduce the **topic**. You can go through the questions to see how much you already know – and you and your teacher can use your answers to see what level you are at, at the start of the topic. You can also see what the big ideas are that you will cover in this topic.

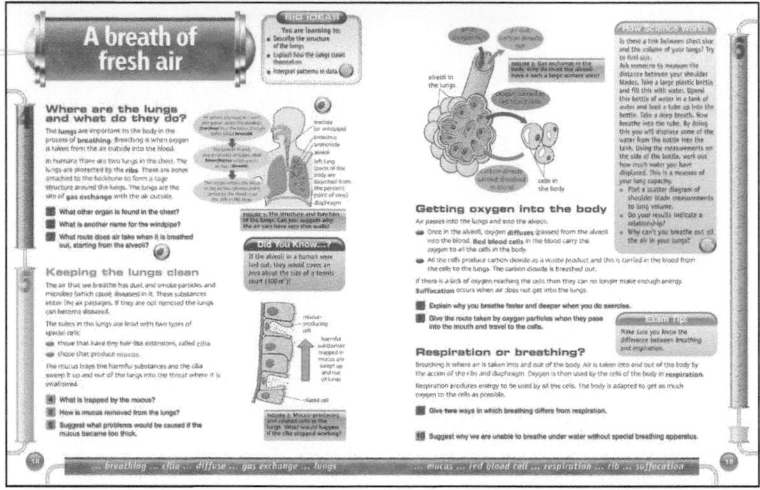

Levelled Lessons

The colour coded levels at the side of the page increase as you progress through the lesson – so you can always see how you are learning new things, gaining new skills and **boosting** your level. Throughout the book, look out for fascinating facts and for hints on how to avoid making common mistakes. The keywords along the bottom can be looked up in the glossary at the back of the book.

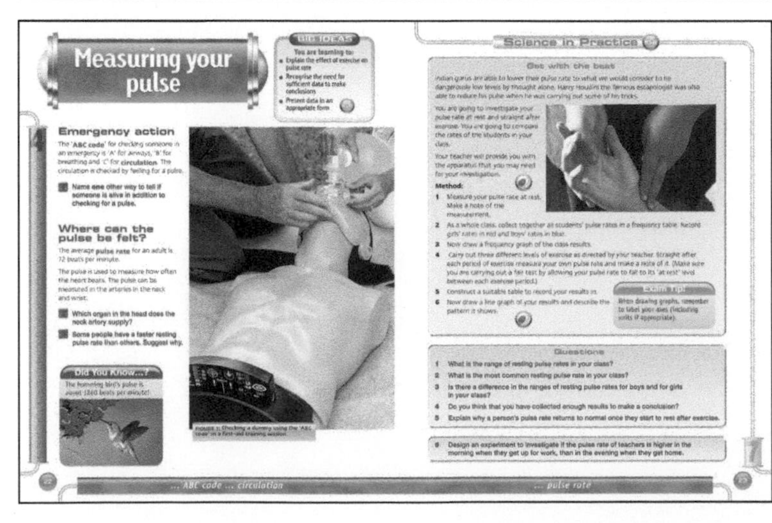

Get Practical

Practicals are what Science is all about! These lessons give you instructions on how to carry out your experiments or investigations, as well as learning about the Science behind it.

Look out for the HSW icons throughout all of the lessons – this is where you are learning about How Science Works.

Welcome to Collins KS3 Science!

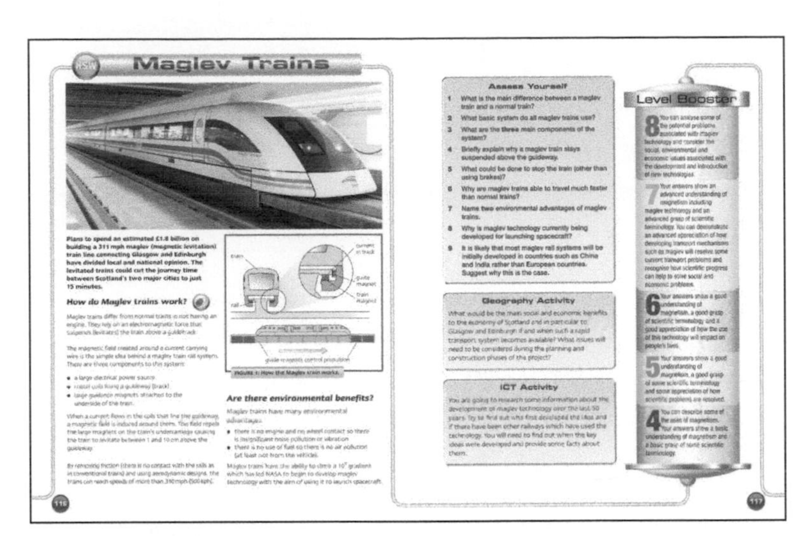

Maglev Trains

Plans to spend an estimated £1.8 billion on building a 311 mph maglev (magnetic levitation) train line connecting Glasgow and Edinburgh have divided local and national opinion. The levitated trains could cut the journey time between Scotland's two major cities to just 15 minutes.

How do Maglev trains work?

Maglev trains differ from normal trains in not having an engine. They rely on an electromagnetic force that suspends (levitates) the train above a guidetrack.

The magnetic field created around a current carrying wire is the simple idea behind a maglev train rail system. There are three components to this system:

- a large electrical power source
- metal coils lining a guideway (track)
- large guidance magnets attached to the underside of the train.

When a current flows in the coils that line the guideway, a magnetic field is induced around them. The field repels the large magnets on the train's undercarriage causing the train to levitate between 1 and 10 cm above the guideway.

By removing friction (there is no contact with the rails as in conventional trains) and using aerodynamic designs, the trains can reach speeds of more than 310 mph (500 kph).

FIGURE 1: How the Maglev train works.

Are there environmental benefits?

Maglev trains have many environmental advantages:

- there is no engine and no wheel contact so there is insignificant noise pollution or vibration
- there is no use of fuel so there is no air pollution (at least not from the vehicle).

Maglev trains have the ability to climb a 10° gradient which has led NASA to begin to develop maglev technology with the aim of using it to launch spacecraft.

Assess Yourself

1. What is the main difference between a maglev train and a normal train?
2. What basic system do all maglev trains use?
3. What are the three main components of the system?
4. Briefly explain why a maglev train stays suspended above the guideway.
5. What could be done to stop the train (other than using brakes)?
6. Why are maglev trains able to travel much faster than normal trains?
7. Name two environmental advantages of maglev trains.
8. Why is maglev technology currently being developed for launching spacecraft?
9. It is likely that most maglev rail systems will be initially developed in countries such as China and India rather than European countries. Suggest why this is the case.

Geography Activity

What would be the main social and economic benefits to the economy of Scotland and in particular to Glasgow and Edinburgh if and when such a rapid transport system becomes available? What issues will need to be considered during the planning and construction phases of the project?

ICT Activity

You are going to research some information about the development of maglev technology over the last 50 years. Try to find out who first developed the idea and if there have been other railways which have used the technology. You will need to find out when the key ideas were developed and provide some facts about them.

Level Booster

8 You can analyse some of the potential problems associated with maglev technology and consider the social, environmental and economic values associated with the development and introduction of new technologies.

7 Your answers show an advanced understanding of magnetism including maglev technology and an advanced grasp of scientific terminology. You can demonstrate an advanced appreciation of how developing transport mechanisms such as maglev will resolve some current transport problems and recognise how scientific progress can help to solve social and economic problems.

6 Your answers show a good understanding of magnetism, a good grasp of scientific terminology and a good appreciation of how the use of this technology will impact on people's lives.

5 Your answers show a good understanding of magnetism, a good grasp of some scientific terminology and some appreciation of how scientific problems are resolved.

4 You can describe some of the uses of magnetism. Your answers show a basic understanding of magnetism and a basic grasp of some scientific terminology.

116 117

Mid-Topic Assessments

About half way through the Topic you get a chance to see how much you've learned already about the big ideas by answering questions on a stimulating article. The Level Booster allows you to see for yourself what level your answers will reach – and what more you would need to add to your answers to go up to the next level. There are also opportunities here to see how the Topic relates to other subjects you are studying.

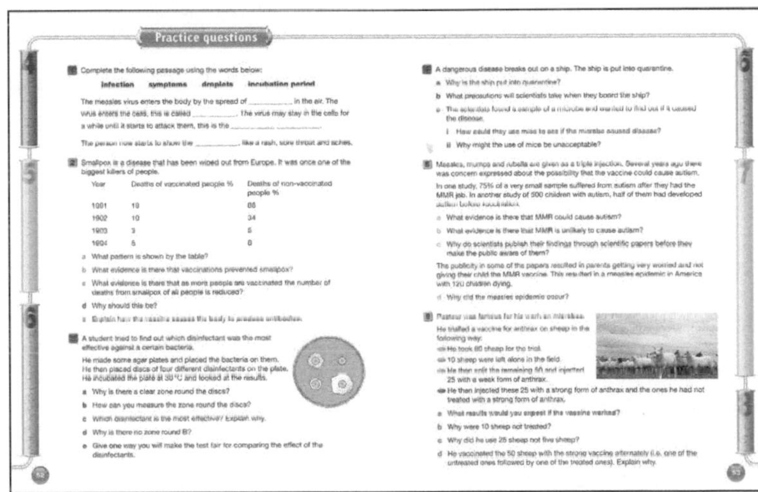

Practice questions

1 Complete the following passage using the words below:

infection symptoms droplets incubation period

The measles virus enters the body by the spread of ________ in the air. The virus enters the cells, this is called ________. The virus may stay in the cells for a while until it starts to attack them, this is the ________ ________.

The person now starts to show the ________, like a rash, sore throat and aches.

2 Smallpox is a disease that has been wiped out from Europe. It was once one of the biggest killers of people.

Year	Deaths of vaccinated people %	Deaths of non-vaccinated people %
1901	19	86
1902	10	34
1903	3	5
1904	5	8

a What pattern is shown by the table?
b What evidence is there that vaccinations prevented smallpox?
c What evidence is there that as more people are vaccinated the number of deaths from smallpox of all people is reduced?
d Why should this be?
e Explain how the vaccine causes the body to produce antibodies.

3 A student tried to find out which disinfectant was the most effective against a certain bacteria.
He made some agar plates and placed the bacteria on them. He then placed discs of four different disinfectants on the plate. He incubated the plate at 30 °C and looked at the results.

a Why is there a clear zone round the discs?
b How can you measure the zone round the discs?
c Which disinfectant is the most effective? Explain why.
d Why is there no zone round B?
e Give one way you will make the test fair for comparing the effect of the disinfectants.

4 A dangerous disease breaks out on a ship. The ship is put into quarantine.

a Why is the ship put into quarantine?
b What precautions will scientists take when they board the ship?
c The scientists found a sample of a microbe and wanted to find out if it caused the disease.
  i How could they use mice to see if the microbe caused disease?
  ii Why might the use of mice be unacceptable?

5 Measles, mumps and rubella are given as a triple injection. Several years ago there was concern expressed about the possibility that the vaccine could cause autism.
In one study, 75% of a very small sample suffered from autism after they had the MMR jab. In another study of 500 children with autism, half of them had developed autism before vaccination.

a What evidence is there that MMR could cause autism?
b What evidence is there that MMR is unlikely to cause autism?
c Why do scientists publish their findings through scientific papers before they make the public aware of them?

The publicity in some of the papers resulted in parents getting very worried and not giving their child the MMR vaccine. This resulted in a measles epidemic in America with 120 children dying.

d Why did the measles epidemic occur?

6 Pasteur was famous for his work on microbes.
He trialled a vaccine for anthrax on sheep in the following way:
- He took 60 sheep for the trial.
- 10 sheep were left alone in the field.
- He then split the remaining 50 and injected 25 with a weak form of anthrax.
- He then injected these 25 with a strong form of anthrax and the ones he had not treated with a strong form of anthrax.

a What results would you expect if the vaccine worked?
b Why were 10 sheep not treated?
c Why did he use 25 sheep not five sheep?
d He vaccinated the 50 sheep with the strong vaccine alternately (i.e. one of the untreated ones followed by one of the treated ones). Explain why.

52 53

Practice Questions

At the end of the Topic you can use these questions to test what you have learned. The questions are all levelled so you can stretch yourself as far as possible!

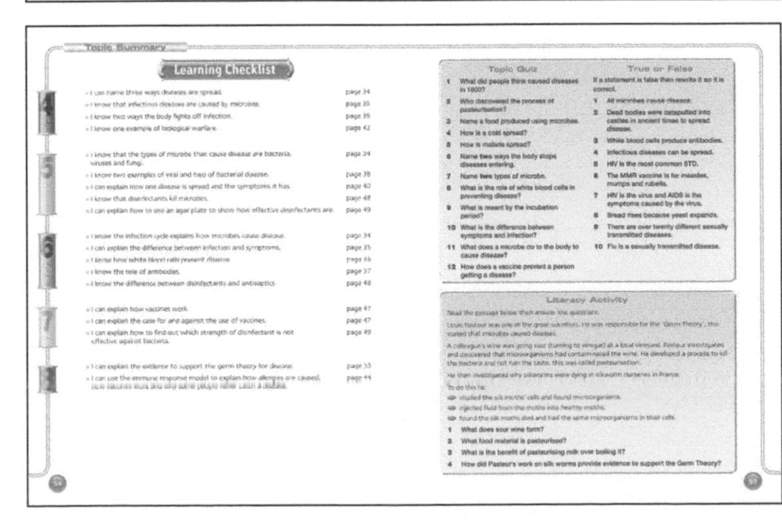

Topic Summary

Learning Checklist

- I can name three ways diseases are spread. page 34
- I know that infectious diseases are caused by microbes. page 35
- I know two ways the body fights off infection. page 35
- I know one example of biological warfare. page 42
- I know that the types of microbe that cause disease are bacteria, viruses and fungi. page 34
- I know two examples of viral and two of bacterial disease. page 38
- I can explain how one disease is spread and the symptoms it has. page 40
- I know that disinfectants kill microbes. page 48
- I can explain how to use an agar plate to show how effective disinfectants are. page 49
- I know the infection cycle explains how microbes cause disease. page 34
- I can explain the difference between infection and symptoms. page 35
- I know how white blood cells prevent disease. page 36
- I know the role of antibodies. page 37
- I know the difference between disinfectants and antiseptics. page 48
- I can explain how vaccines work. page 47
- I can explain the case for and against the use of vaccines. page 47
- I can explain how to find out which strength of disinfectant is not effective against bacteria. page 49
- I can explain the evidence to support the germ theory for disease. page 33
- I can use the immune response model to explain how allergies are caused, how vaccines work and why some people never catch a disease. page 44

Topic Quiz

1. What did people think caused diseases in 1800?
2. Who discovered the process of pasteurisation?
3. Name a food produced using microbes.
4. How is a cold spread?
5. How is malaria spread?
6. Name two ways the body stops diseases entering.
7. Name two types of microbe.
8. What is the role of white blood cells in preventing disease?
9. What is meant by the incubation period?
10. What is the difference between symptoms and infection?
11. What does a microbe do to the body to cause disease?
12. How does a vaccine prevent a person getting a disease?

True or False

If a statement is false then rewrite it so it is correct.

1. All microbes cause disease.
2. Dead bodies were catapulted into castles in ancient times to spread disease.
3. White blood cells produce antibodies.
4. Infectious diseases can be spread.
5. HIV is the most common STD.
6. The MMR vaccine is for measles, mumps and rubella.
7. HIV is the virus and AIDS is the symptoms caused by the virus.
8. Bread rises because yeast expands.
9. There are over twenty different sexually transmitted diseases.
10. Flu is a sexually transmitted disease.

Literacy Activity

Read the passage below then answer the questions.

Louis Pasteur was one of the great scientists. He was responsible for the 'Germ Theory', this stated that microbes caused diseases.

A colleague's wine was going sour (turning to vinegar) at a local vineyard. Pasteur investigated and discovered that microorganisms had contaminated the wine. He developed a process to kill the bacteria and not ruin the taste, this was called pasteurisation.

He then investigated why silkworms were dying in silkworm nurseries in France.

To do this he:
- studied the silk moths' cells and found microorganisms.
- injected fluid from the moths into healthy moths.
- found the silk moths died and had the same microorganisms in their cells.

1. What does sour wine form?
2. What food material is pasteurised?
3. What is the benefit of pasteurising milk over boiling it?
4. How did Pasteur's work on silk worms provide evidence to support the Germ Theory?

Topic Summary

Check what you have learned in the levelled Learning checklist - you can then see at a glance what you might need to go back and revise again. There's a fun Quiz and an activity linked in to another subject so you can be really sure you've got the Topic covered before you move on to the next one.

You are what you eat

This is a food pyramid and it shows you the amount of different kinds of food that should be included in a healthy, balanced diet.

- At the base of the pyramid are bread, cereal, rice and pasta – a healthy diet should have more of these kinds of food than any other.
- Next is fruit and vegetables – a healthy diet should include three to five portions per day.
- Then there is milk, cheese, yoghurt, meat, fish, beans, eggs and nuts. People who follow a vegetarian diet need to ensure that they eat more of beans and nuts to make sure that their diet is not unbalanced due to a lack of meat.
- At the top of the pyramid are fats, oils and sweets. Even a healthy diet can include treats – we need fats and sugars to give us energy – but we must make sure that we only have small portions of these and not too often!

KEEPING HEALTHY

BIG IDEAS

By the end of this unit you will be able to describe the role of nutrients and of the digestive and circulatory systems. You will be able to explain how the process of respiration involves cells and the transfer of energy. You will be able to use evidence from experiments and elsewhere to develop explanations.

Collins Café Menu

Breakfast

Wholemeal toast with butter

1 poached egg

Glass of fresh orange juice

Morning snack

1 apple

Lunch

A wholemeal bread sandwich, made with chicken, mayonnaise and tomato

1 fruit yoghurt

1 bag of crisps

Afternoon snack

1 small chocolate bar

Dinner

Spaghetti bolognaise made with fried minced meat, tomatoes, garlic and onions

Salad with lettuce and cucumber

Evening snack

Glass of milk

Handful of nuts

Recommendations for a healthy diet are as follows.

- Eat a varied diet with a wide range of foods.
- Eat more wholegrain starchy carbohydrates. Starchy foods should make up about a third of the food you eat. They are a good source of energy, fibre, minerals and vitamin B.
- Eat more fruit and vegetables.
- Reduce the amount of salt that you eat (think about looking at the labels on ready meals – you may be surprised how much salt they contain!).
- Eat regular meals.
- Control the portion size.

What do you know?

1 a Name **three** foods rich in starch.
  b Where are starchy foods in the food pyramid?

2 a Name **three** fruits.
  b Where are fruits in the food pyramid?

3 a Name **two** foods that are rich in proteins.
  b Where are foods rich in protein in the pyramid?

4 a Name **one** food that is rich in fat.
  b Where are foods rich in fat in the pyramid?

5 Which type of food is least expensive to buy?

6 Suggest which foods form the basic diet of people in the developing world.

7 Explain why the pyramid has been constructed in this way.

8 Have a look at the *Collins Café Menu*. Using the food pyramid as a guide, explain why this menu might or might not be described as a balanced daily diet.

9 Construct your own pyramid to show what you ate yesterday.

10 What changes can you make to your diet to make it healthier?

A balanced diet

BIG IDEAS

You are learning to:

- Identify examples of each type of food
- Explain the importance of a balanced diet
- Use information to make decisions about lifestyle

A balanced diet

A **balanced diet** contains different **food types** in the correct amounts.

If a person eats an **unbalanced diet**, they lack one or more of the food types their body needs to stay healthy. They will suffer from **malnutrition**.

1 Give **one** example of a food that you eat that is rich in:
a protein **b** fat **c** minerals.

Did You Know...?

Bananas are eaten by sports players as they are rich in carbohydrates, vitamins and minerals and easy to digest.

How Science Works

Food types

2 Suggest why a young baby needs a lot of calcium.

3 Give **two** types of food in cheese.

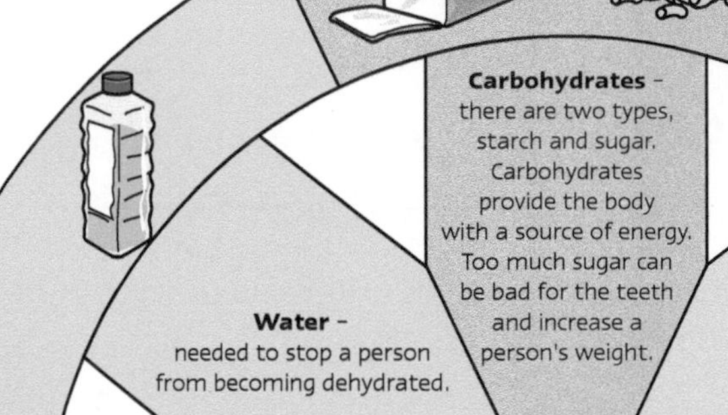

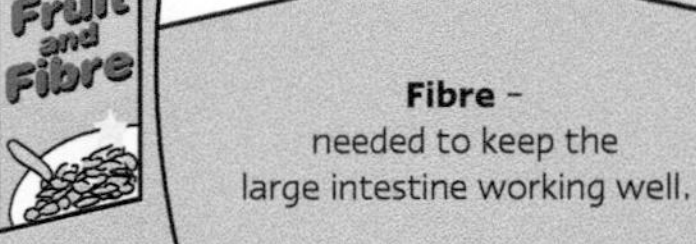

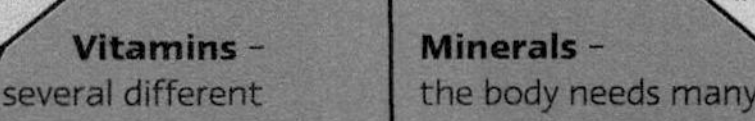

FIGURE 1: Examples of foods of each food type and their uses in the body.

... balanced diet ... constipation ... food type

The importance of fibre

Fibre mainly consists of a carbohydrate called cellulose. (It forms the cell walls in plants.) It cannot be digested by the body and adds bulk to the food in the intestines. This helps the intestines push food through using the muscles in the intestine wall. If a person has a low fibre intake they can suffer from **constipation**. This is when the faeces, which are usually soft, become hard because of a lack of fibre. To pass hard faeces, a person with constipation needs to strain and this may cause damage to their intestinal wall. This damage increases the risk of them suffering from bowel cancer.

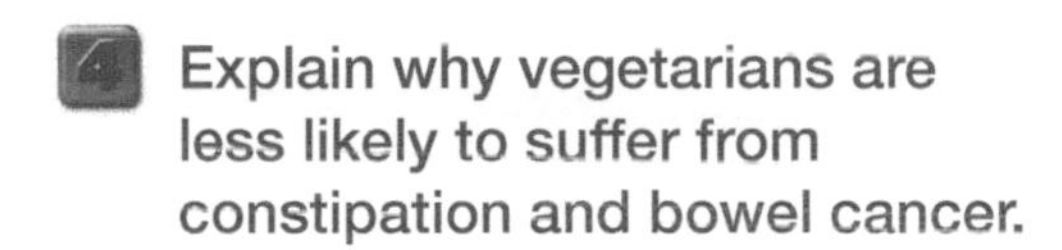

4 Explain why vegetarians are less likely to suffer from constipation and bowel cancer.

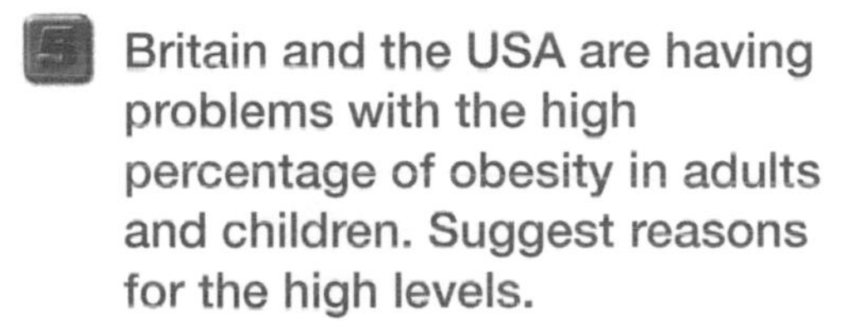

5 Britain and the USA are having problems with the high percentage of obesity in adults and children. Suggest reasons for the high levels.

How Science Works

Food Labels

Food labels indicate the nutritional value of the food. The 'traffic light' system has been introduced to help show you the levels of fat, saturated fats, sugar and salt. Red indicates the content of the food type is high, amber indicates the content is medium and green indicates the content is low. The amount of fat, saturated fat, sugar and salt is given in grams per portion of the food.

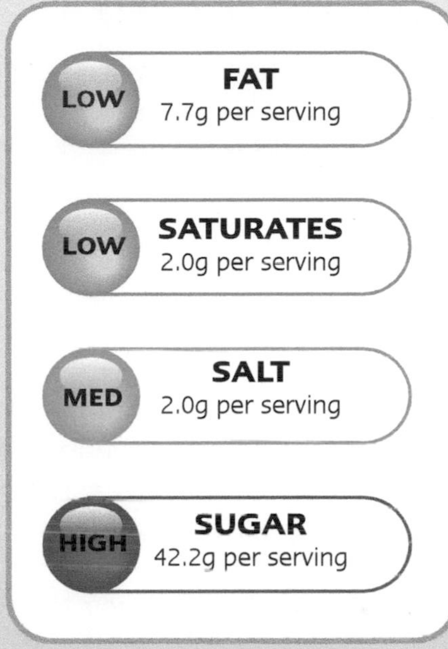

A food label has recently been designed that shows how long fresh food can be stored before it needs to be thrown away. When the food is opened an ink line appears on the special label that grows longer by the day, eventually becoming full-length which indicates that the food is no longer good to eat.

1 How can a traffic light system help you choose food for a healthy diet?

2 Why is a traffic light system better than just having the nutritional value of the food?

3 How could a food label which indicates the shelf life of food benefit our environment?

HSW

Vitamin supplements

Vitamin and mineral **supplements** can be bought at chemist shops. Some people take these supplements instead of eating fruit and vegetables. In some cases they take too many and this can lead to side-effects.

6 Suggest why it is better to eat fresh fruit and vegetables as a source of vitamin C rather than to take tablets.

FIGURE 2: Why is it important to eat lots of fruit and vegetables?

7 Junk food is thought to have caused many dietary problems. Explain what junk food is and explain in detail the problems it has caused.

Is my diet OK?

BIG IDEAS

You are learning to:

- Explain why it is important to eat a healthy diet
- Evaluate the danger of eating salt
- Explain why it is important to eat breakfast

Five-a-day 

Your five portions of **fruit** and **vegetables** can be taken in different ways:

- as fresh fruit
- as juices
- as canned, dried or frozen
- in salad.

They provide vitamins, fibre and minerals. There is good evidence that they cut the risk of diseases such as heart disease and cancer.

1. Which vitamin is in fruit?
2. Which part of an apple has a lot of fibre in?

Fish for the brain

Fish is an excellent source of protein and contains many minerals and vitamins. Oily fish, such as mackerel and salmon, are rich in **omega-3 fatty acids**. These fatty acids help to keep a person's heart healthy and are an important nutrient for the brain.

3. Why does a person need protein in their diet?
4. Why are omega-3 fatty acids important for the diet?

The Golden Rules for healthy eating

- **Eat five portions of fruit or vegetables a day.**

- **Eat a lot of starchy foods.**

- **Choose low-fat versions of foods.**

- **Cut down on foods which are high in salt.**

- **Cut down on sugary foods.**

- **Eat more fish.**

- **Do not skip breakfast.**

- **Drink plenty of water.**

... blood pressure ... fruit ... omega-3 fatty acids

The dangers of too much salt

The body needs **salt** to keep working. However:

- too little salt can stop the brain, heart and muscles working properly
- too much salt causes too much water to be kept by the body. This leads to swelling of the hands, feet and sometimes the abdomen. It also causes an increase in **blood pressure** that increases the risk of stroke and heart attack.

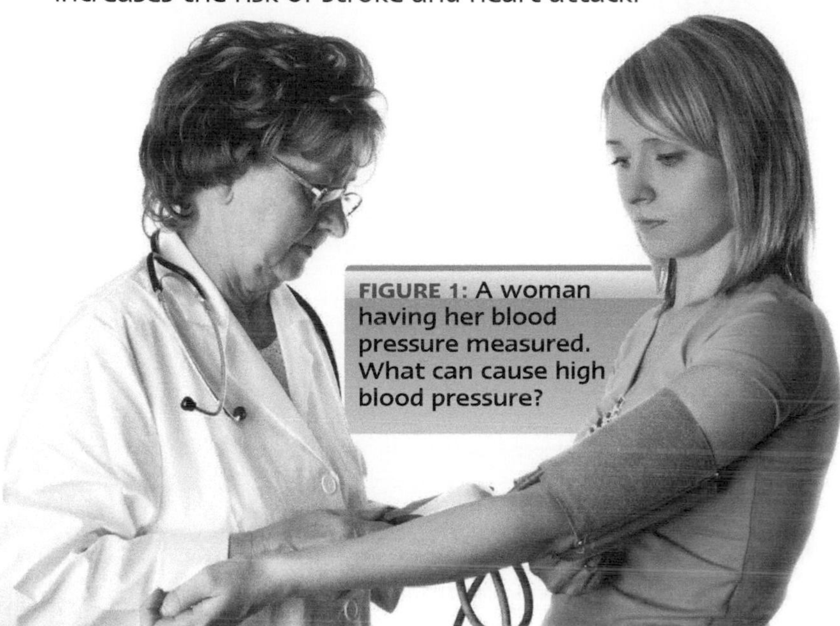

FIGURE 1: A woman having her blood pressure measured. What can cause high blood pressure?

Did You Know...?

In ancient times salt was once as precious as gold. It was used in foods as a preservative and as a flavouring and also as a dye, bleach and to soften leather. In fact it was so precious that soldiers were paid in salt, and this is where our word 'salary' comes from today.

How Science Works

Planning a healthy diet is not easy! You need to look at what you like to eat and change those things which are not healthy.

- Plan a whole day's diet where you can eat whatever you want, and forget about being healthy.
- Plan a whole day's diet following the Golden Rules.
- Compare the diets and discuss how changes can be made.

HSW

5 Why does too much salt cause:
a swelling of the feet
b an increase in weight?

6 Suggest how a person could change their diet in order to lower their salt intake.

Eat breakfast!

Breakfast gives the body an energy boost at the start of the day. If a person misses breakfast they do not necessarily lose weight – in fact research has shown eating breakfast helps to maintain a person's weight, possibly because they eat fewer snacks later on.

FIGURE 2: Why is it important to eat breakfast?

7 Why is it good to eat starchy foods and fruit for breakfast?

8 Porridge contains a 'slow-release' carbohydrate. Suggest why this is an advantage.

9 Explain, in terms of energy consumption, why having breakfast is less likely to make you gain weight than not having breakfast.

Eating food

BIG IDEAS

You are learning to:
- Describe what happens as food passes through the digestive system
- Explain what is meant by digestion
- Explain the role of enzymes

The map of the digestive system

The food a person eats must be broken down into smaller pieces. This is carried out in the **digestive system**. The food is pushed through the digestive system by muscles that contract.

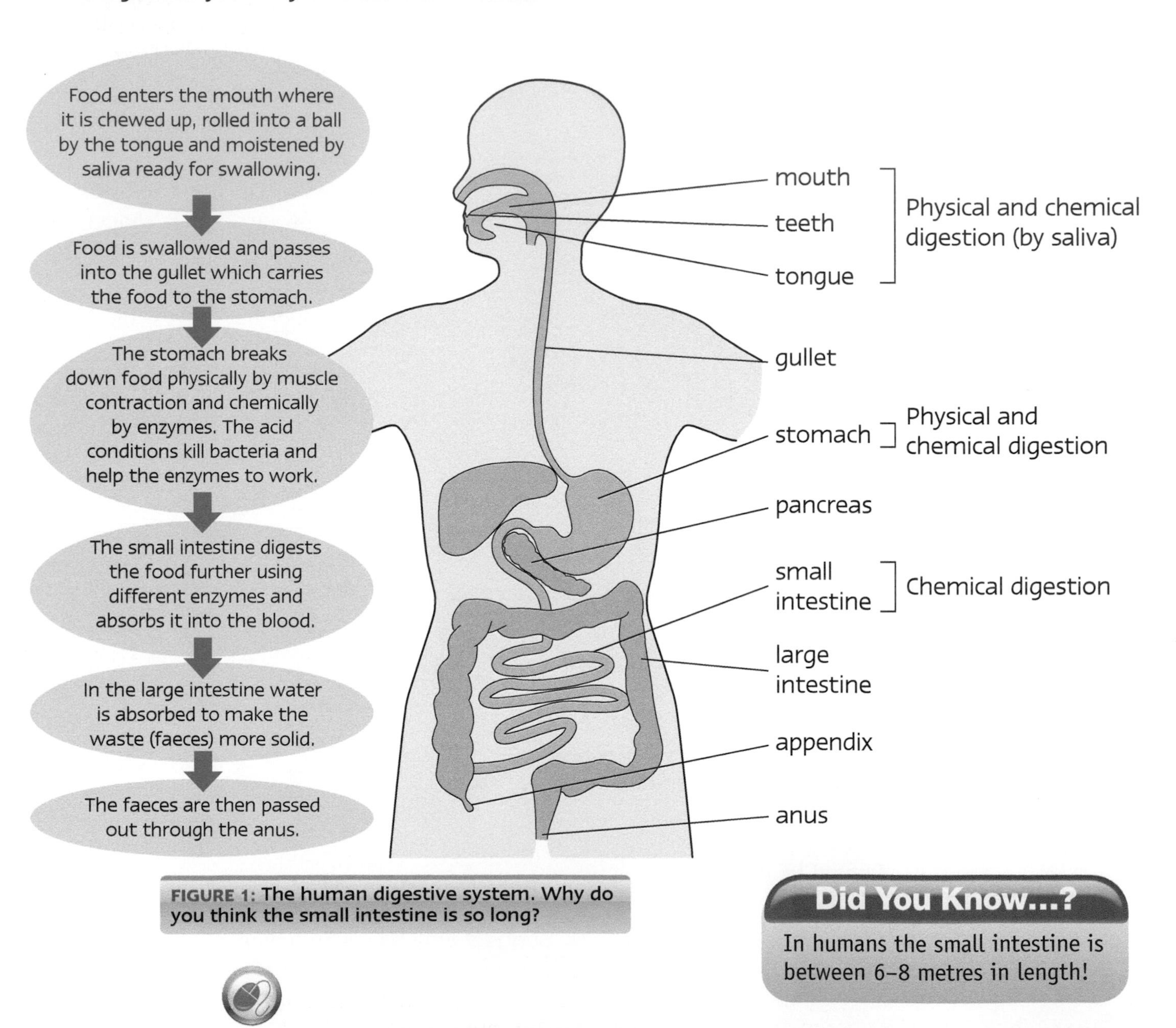

FIGURE 1: The human digestive system. Why do you think the small intestine is so long?

Did You Know...?

In humans the small intestine is between 6–8 metres in length!

... amino acid ... chemically ... digestion ... digestive system

Digestion

- Food is broken down **physically** by the teeth in the mouth and by muscles lining the digestive system.
- Food is broken down **chemically** by substances called **enzymes**. These are like chemical 'scissors' that cut up large food **molecules** into smaller food molecules. This process is called **digestion**. As a result of digestion food molecules are small enough to be absorbed into the blood in the small intestine.

Food that is not digested is passed out of the body. It is called **faeces**.

What happens to a cheese roll?

A cheese roll contains large molecules of protein and carbohydrates (starch). When the roll is eaten and digested, the starch is broken down into smaller molecules of sugar and the protein is broken down into smaller molecules called **amino acids**. These molecules are small enough to pass into the blood.

1. List the organs (in the correct order) that food passes through.
2. Give **two** reasons why conditions in the stomach are acidic.
3. What does 'digestion' mean?
4. Why is food broken down into smaller molecules during digestion?
5. Which type of food helps food pass through the intestine?
6. Explain what happens to a ham roll as it passes through your digestive system.

7. Explain how the model of matching scissors to enzymes helps explain how enzymes work. How does the model need to be modified to explain the digestion of proteins and starch?
8. A person with cystic fibrosis releases few enzymes into their small intestine. Explain what problems this causes.

Did You Know...?

It takes quite a long time for your food to pass through your digestive system.

0 hours	food is chewed and swallowed
1 hours	food is churned with acids and enzymes in the stomach
2 hours	partially-digested food passes into the small intestine for further digestion and the start of absorption into the blood
6 hours	undigested food passes into the large intestine and water is taken out of it and passes back into the blood
10 hours	the leftovers collect in the rectum
16–24hrs	the faeces pass out of the body

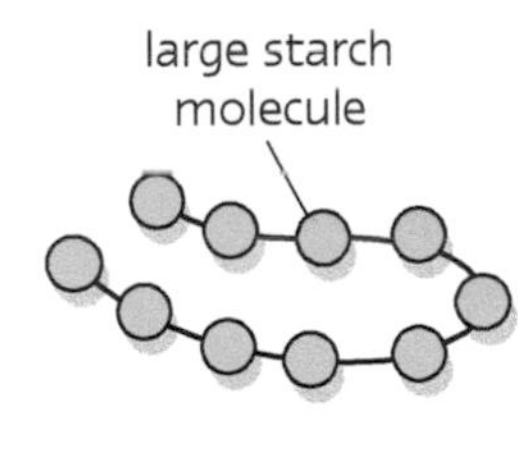

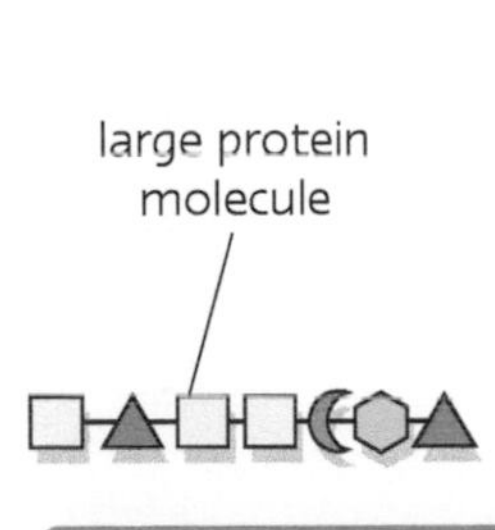

FIGURE 2: The fate of a cheese roll!

... *faeces* ... *enzyme* ... *molecule* ... *physically*

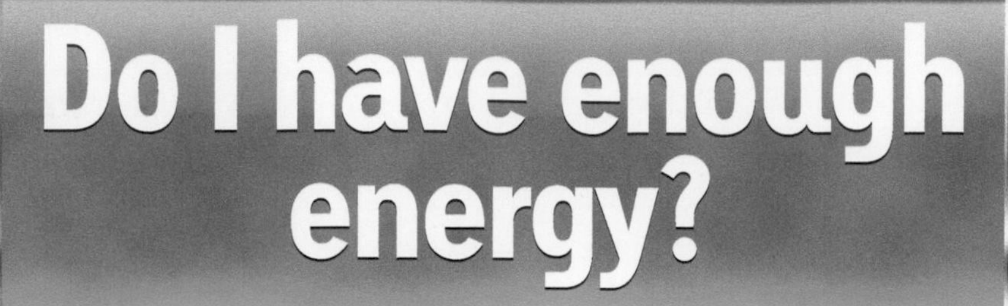

Do I have enough energy?

BIG IDEAS

You are learning to:
- Describe how the body uses energy
- Explain what is meant by respiration
- Analyse the energy requirements of different people

Using energy

Energy comes from our food. This energy is important as it allows the body to:

- grow
- repair
- keep warm
- move (by muscles contracting)
- feel (nerves).

Did You Know...?

The energy value in one bag of crisps is the same as eating one banana *and* a small portion of fish and chips!

1. When you are walking on a cold day what is some of the energy in your body being used for?
2. When you are asleep what is some of the energy in your body being used for?

FIGURE 1: Even when you are relaxing, your body needs energy.

Energy use changes with age

The amount of energy a person needs varies in their lifetime. The table on the next page shows how much energy is needed at different stages in a person's life.

3. Suggest why a 16-year-old boy or girl needs more energy in a day than a baby.
4. Suggest why a 16-year-old boy needs more energy in a day than a 16-year-old girl.
5. Suggest why a labourer needs more energy in a day than an office worker.

FIGURE 2: Have you heard of the saying 'working up an appetite'?

... carbon dioxide ... glucose ... oxygen

Person	Energy used in a day (kJ)
baby	3300
8-year-old (boy or girl)	6500
16-year-old boy	12 000

Person	Energy used in a day (kJ)
16-year-old girl	8800
average adult man	11 500
average adult woman	9000

Person	Energy used in a day (kJ)
office worker (man)	10 500
labourer (manual worker) (man)	15 500
pregnant woman	10 000

Respiration

Respiration is the transformation of stored chemical energy in food to useful energy for the body.

Read the definition of respiration. The useful energy is used to fuel reactions that take place in all living cells.

Oxygen and **glucose** (a type of sugar) needed for respiration are carried to all the cells in the blood. The substances made (**products**) in the **reaction** are **carbon dioxide** and water.

oxygen + glucose → carbon dioxide + water + (energy)

6 In which organ does the body take in oxygen and remove carbon dioxide?

7 How is oxygen transported to all the cells?

8 Explain why a manual worker will be less likely to put on weight if he snacks than an office worker.

Middle-aged spread

As a person grows older the basic rate of respiration in their body tends to slow down. Older people are on average less active and need less energy for growth. If they were very active when they were young and then stop being active they put on weight. Many sports people gain weight after they have given up sport at a competitive level.

9 People often put on weight in middle age. We call this 'middle-aged spread'. Suggest why this happens and how it could be avoided.

10 Explain why crocodiles and snakes only need to eat every few days, whereas humans need to eat regularly. Relate your answer to the energy requirement.

... product ... reaction ... respiration

Free health MOT

***HealthFirst* is offering a free health MOT to all its members this Saturday. Just as the oil in a car is checked regularly, so it is important to make sure that all the parts of your body are working properly.**

You and your family's medical history. We will ask you some simple questions so that we can build up a picture of your health history and lifestyle.

Weight. First we will compare your weight with our standard charts to see if you are underweight, an acceptable weight, overweight or obese (very overweight). Obesity increases your risk of suffering from heart disease, diabetes and joint problems.

Heart rate. Then we will measure your heart rate when you are at rest and display it on the screen of our heart rate monitor. It will produce a picture like the one shown below.

We will then measure your heart rate when you are exercising. Any irregularities in the pattern on the screen may indicate faulty valves in the heart and poor blood flow to the heart muscle.

Blood. We will then take a small sample of blood and count the number of blood cells and analyse the plasma for glucose, blood proteins and cholesterol. We will use centrifugation and chemical tests to do this.

From our results we can tell if other organs in your body are working properly.

There's nothing to worry about! Here's a completed *HealthFirst* MOT for you to look at.

Centrifugation (very fast spinning) and chemical tests in test tubes are used to analyse a blood sample.

A health check, can sort out problems before they become serious so don't miss out on this chance to check that your body is in top condition!

The spikes show the contractions of the heart. They should occur regularly and have the same repeating pattern.

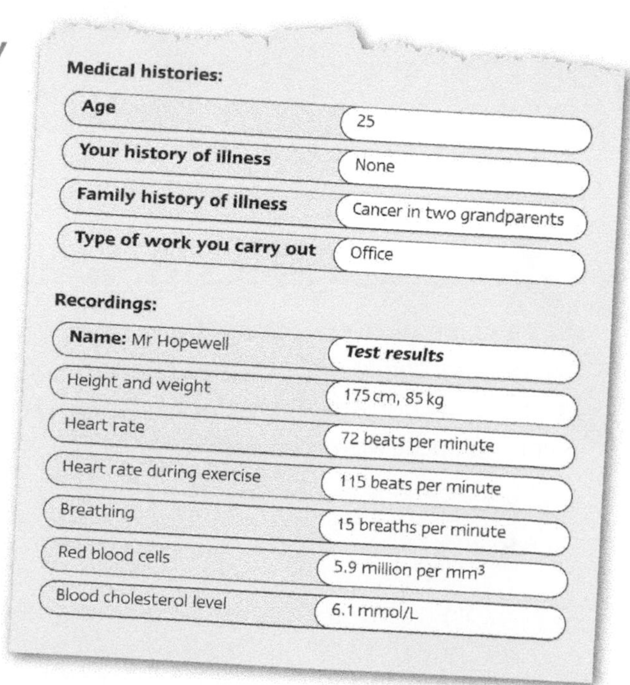

Medical histories:

Age	25
Your history of illness	None
Family history of illness	Cancer in two grandparents
Type of work you carry out	Office

Recordings:

Name: Mr Hopewell	***Test results***
Height and weight	175 cm, 85 kg
Heart rate	72 beats per minute
Heart rate during exercise	115 beats per minute
Breathing	15 breaths per minute
Red blood cells	5.9 million per mm^3
Blood cholesterol level	6.1 mmol/L

Assess Yourself

1. What is the purpose of a health check?
2. Which tests are carried out on the blood?
3. Stuart is 22 years old and his health check indicates that he is obese.
   - **a** Which types of food would you recommend to give him a healthy diet?
   - **b** What problems might Stuart experience if he does not lose weight?
4. If John has a low red blood cell count, what would the doctor recommend he eats?
5. Mr Hopewell is an office worker. Why should he be careful about the energy intake from his diet?
6. Explain why a person is only expected to carry out light exercise when the heart rate is measured.
7. Explain why red blood cells collect at the bottom of a tube when centrifuged.
8. Explain the use made of the data collected about a person's blood.
9. It may soon be possible to attach yourself to a device on a computer and have the information about your body sent to a health clinic to receive an online diagnosis. Do you think this is a good idea? Give reasons for and against this method.

ICT Activity

1. Prepare a leaflet on the computer to advertise the health MOT to people aged 25–30 years. Think how to make your leaflet exciting and informative.
2. Prepare a daily light exercise programme for a person who needs to lose weight slowly.

Level Booster

8 You can give a reasoned argument of the case for or against the use of an online health check.

7 You can explain the use made of the data collected about a person's blood.

6 You can explain the procedures used to collect data on a person's blood and the use made of the information to diagnose the health of a person.

5 You can explain the problems caused by being obese.

4 You can describe the basic content of a healthy diet and indicate the food types that need to be eaten to improve a person's health.

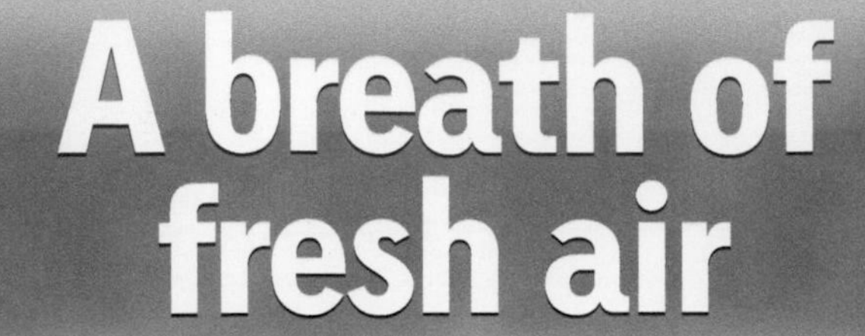

A breath of fresh air

BIG IDEAS

You are learning to:

- Describe the structure of the lungs
- Explain how the lungs clean themselves
- Interpret patterns in data 

Where are the lungs and what do they do?

The **lungs** are important to the body in the process of **breathing**. Breathing is when oxygen is taken from the air outside into the blood.

In humans there are two lungs in the chest. The lungs are protected by the **ribs**. These are bones attached to the backbone to form a cage structure around the lungs. The lungs are the site of **gas exchange** with the air outside.

1. What other organ is found in the chest?
2. What is another name for the windpipe?
3. What route does air take when it is breathed out, starting from the alveoli?

Air enters the nose or mouth and passes down the windpipe (**trachea**) into the lungs through tubes called **bronchi**.

The bronchi branch into a network of tubes called **bronchioles** which end in air sacs (**alveoli**).

The oxygen enters the blood in the air sacs (alveoli) and is carried by the blood to all the cells in the body.

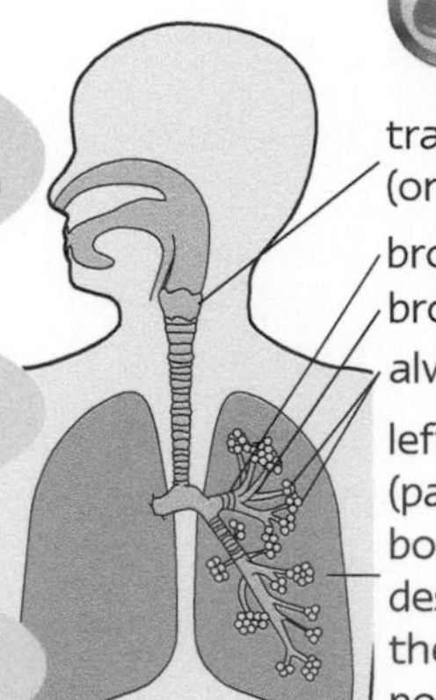

FIGURE 1: **The structure and function of the lungs. Can you suggest why the air sacs have very thin walls?**

Did You Know...?

If the alveoli in a human were laid out, they would cover an area about the size of a tennis court ($100\ m^2$)!

Keeping the lungs clean

The air that we breathe has dust and smoke particles and microbes (which cause diseases) in it. These substances enter the air passages. If they are not removed the lungs can become diseased.

The tubes in the lungs are lined with two types of special cells:

- those that have tiny hair-like extensions, called **cilia**
- those that produce **mucus**.

The mucus traps the harmful substances and the cilia sweep it up and out of the lungs into the throat where it is swallowed.

4. What is trapped by the mucus?
5. How is mucus removed from the lungs?
6. Suggest what problems would be caused if the mucus became too thick.

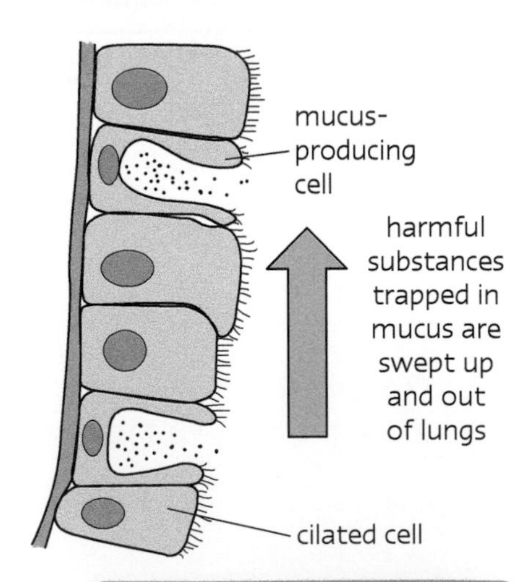

FIGURE 2: **Mucus-producing and ciliated cells in the lungs. What would happen if the cilia stopped working?**

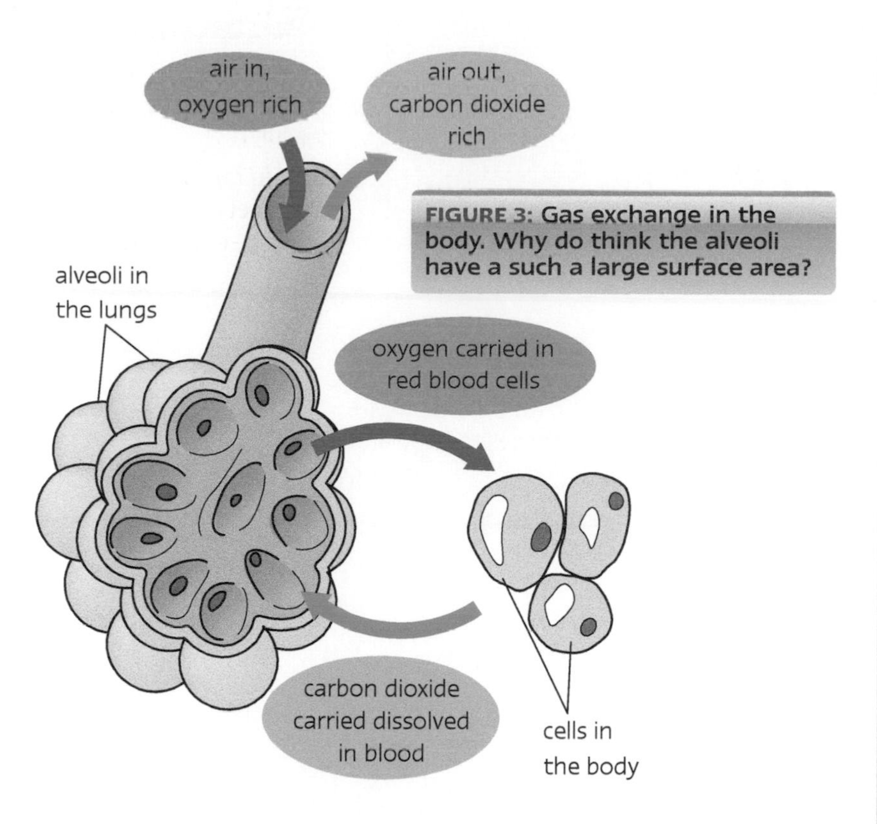

FIGURE 3: Gas exchange in the body. Why do think the alveoli have a such a large surface area?

How Science Works

Is there a link between chest size and the volume of your lungs? Try to find out.
Ask someone to measure the distance between your shoulder blades. Take a large plastic bottle and fill this with water. Upend this bottle of water in a tank of water and feed a tube up into the bottle. Take a deep breath. Now breathe into the tube. By doing this you will displace some of the water from the bottle into the tank. Using the measurements on the side of the bottle, work out how much water you have displaced. This is a measure of your lung capacity.

- Plot a scatter diagram of shoulder blade measurements to lung volume.
- Do your results indicate a relationship?
- Why can't you breathe out all the air in your lungs?

HSW

Getting oxygen into the body

Air passes into the lungs and into the alveoli.

- Once in the alveoli, oxygen **diffuses** (passes) from the alveoli into the blood. **Red blood cells** in the blood carry the oxygen to all the cells in the body.
- All the cells produce carbon dioxide as a waste product and this is carried in the blood from the cells to the lungs. The carbon dioxide is breathed out.

If there is a lack of oxygen reaching the cells then they can no longer make enough energy. **Suffocation** occurs when air does not get into the lungs.

7 Explain why you breathe faster and deeper when you do exercise.

8 Give the route taken by oxygen particles when they pass into the mouth and travel to the cells.

Exam Tip!

Make sure you know the difference between *breathing* and *respiration*.

Respiration or breathing?

Breathing is where air is taken into and out of the body. Air is taken into and out of the body by the action of the ribs and diaphragm. Oxygen is then used by the cells of the body in **respiration**.

Respiration produces energy to be used by all the cells. The body is adapted to get as much oxygen to the cells as possible.

9 Give **two** ways in which breathing differs from respiration.

10 Suggest why we are unable to breathe under water without special breathing apparatus.

A healthy heart

BIG IDEAS

You are learning to:

- Describe the structure of the heart
- Explain how disease affects the heart

4

What does the heart do?

The **heart** pumps **blood** around the body. The heart is made mainly of **muscle** tissue.

The blood is carried in tubes, called **vessels**, to the organs and tissues in the body (like the water supply to the radiators in a house).

- A blood vessel that takes blood away from the heart is called an **artery**. An artery has a **pulse**.
- A blood vessel that takes blood back to the heart is called a **vein**. A vein does not have a pulse.

1 Look at the underside of your arm. What type of blood vessels are the blue ones?

2 Place two fingers on the underside of your wrist on the thumb side. What type of blood vessel can you feel that has a pulse?

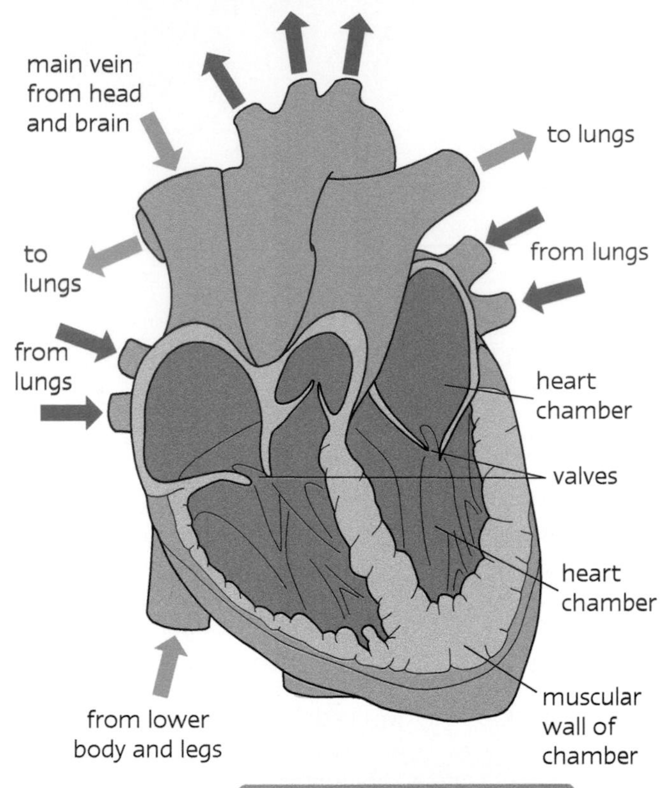

FIGURE 1: The heart. Why do you think the heart has thick walls of muscle?

5

Role of the blood

Blood is made up of a liquid part called **plasma** and red and white blood cells. The blood transports material around the body. It carries oxygen (in red blood cells), glucose and other food molecules to all the cells in the body.

Blood carries carbon dioxide away from the cells to the lungs.

3 Name **two** things the blood transports.

4 What is the liquid part of blood called?

5 Which blood cells carry oxygen?

Did You Know...?

When a person is sitting down their heart beats about 60–75 times each minute, but after lots of exercise it beats at about 130 times a minute.

... artery ... blood ... clot ... coronary artery ... fatty deposit

Oxygen and the heart

The heart's muscle tissue must be provided with a good supply of blood to bring oxygen and glucose to it so that it can contract.

The blood vessel that brings blood to the heart muscle tissue is called the **coronary artery**. This artery can become narrowed by a build up of **fatty deposits** inside it. The blood cannot flow so well through the narrow regions and sometimes blood **clots** are trapped and block the coronary artery, causing a heart attack.

6 Suggest why a person with a large heart has a slower pulse rate than someone with a smaller heart.

7 Why is heart muscle unable to contract properly if the coronary artery becomes blocked?

How Science Works

Olympic long distance races were often dominated by athletes from Kenya. These athletes trained in the mountains in their country. Scientists found their blood contained more red blood cells than athletes who didn't train at height. Red blood cells carry oxygen to the cells. As a result, athletes sometimes travel to mountainous areas as part of their training programme. When the body is first exposed to high altitude, it responds to the lower oxygen levels in the air by increasing its breathing rate and heart rate to get more oxygen into the body.

After a couple of weeks the athlete becomes acclimatised to the altitude. This means their body has made more red blood cells and its capillary network has increased. These changes make the body better at getting oxygen to all its cells.

When the athlete returns to a lower altitude, the body is still able to get the optimum amount of oxygen into its cells for a short period of time.

1 Why would you feel tired if you were walking at heights of 2000 m?

2 Why do mountaineers tackling some of the highest peaks acclimatise at a base camp for several weeks?

3 Why does an increase in red blood cells help athletes running 10 000 m and marathons?

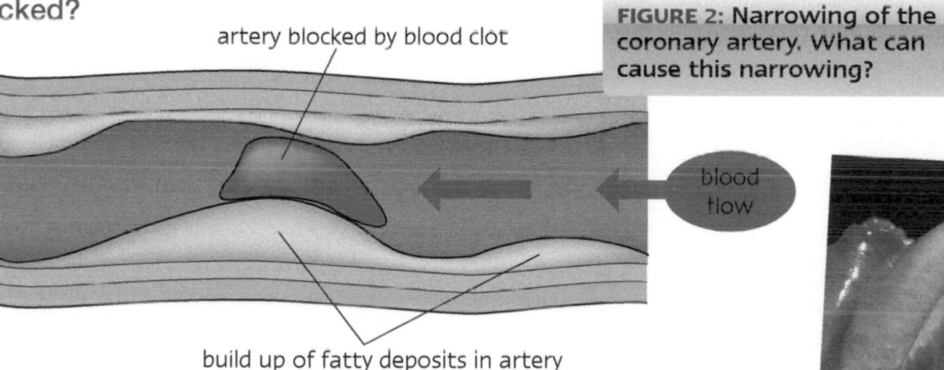

FIGURE 2: Narrowing of the coronary artery. What can cause this narrowing?

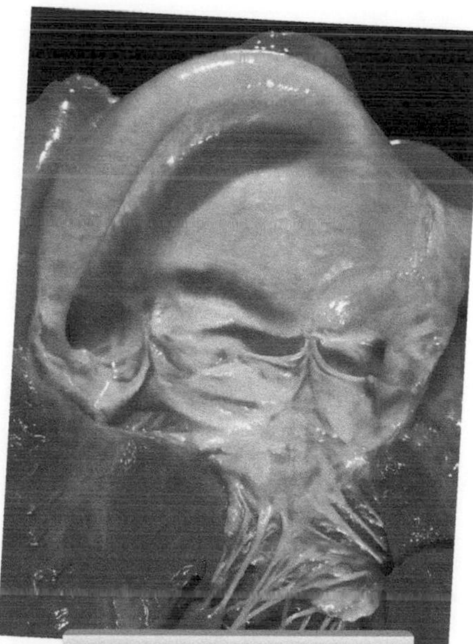

FIGURE 3: Heart valves. What is their function?

Preventing back-flow of blood in the heart

The heart's muscular chambers contract to force the blood through into the next chamber in one direction. Heart **valves** between the chambers stop the blood from flowing backwards (see Figure 1).

8 Explain how the action of the heart muscles and valves ensure blood flows though the heart in one direction.

Measuring your pulse

BIG IDEAS

You are learning to:
- Explain the effect of exercise on pulse rate
- Recognise the need for sufficient data to make conclusions
- Present data in an appropriate form 

4

Emergency action

The '**ABC code**' for checking someone in an emergency is 'A' for airways, 'B' for breathing and 'C' for **circulation**. The circulation is checked by feeling for a pulse.

1. Name **one** other way to tell if someone is alive in addition to checking for a pulse.

Where can the pulse be felt?

The average **pulse rate** for an adult is 72 beats per minute.

The pulse is used to measure how often the heart beats. The pulse can be measured in the arteries in the neck and wrist.

2. Which organ in the head does the neck artery supply?
3. Some people have a faster resting pulse rate than others. Suggest why.

Did You Know...?

The humming bird's pulse is about 1260 beats per minute!

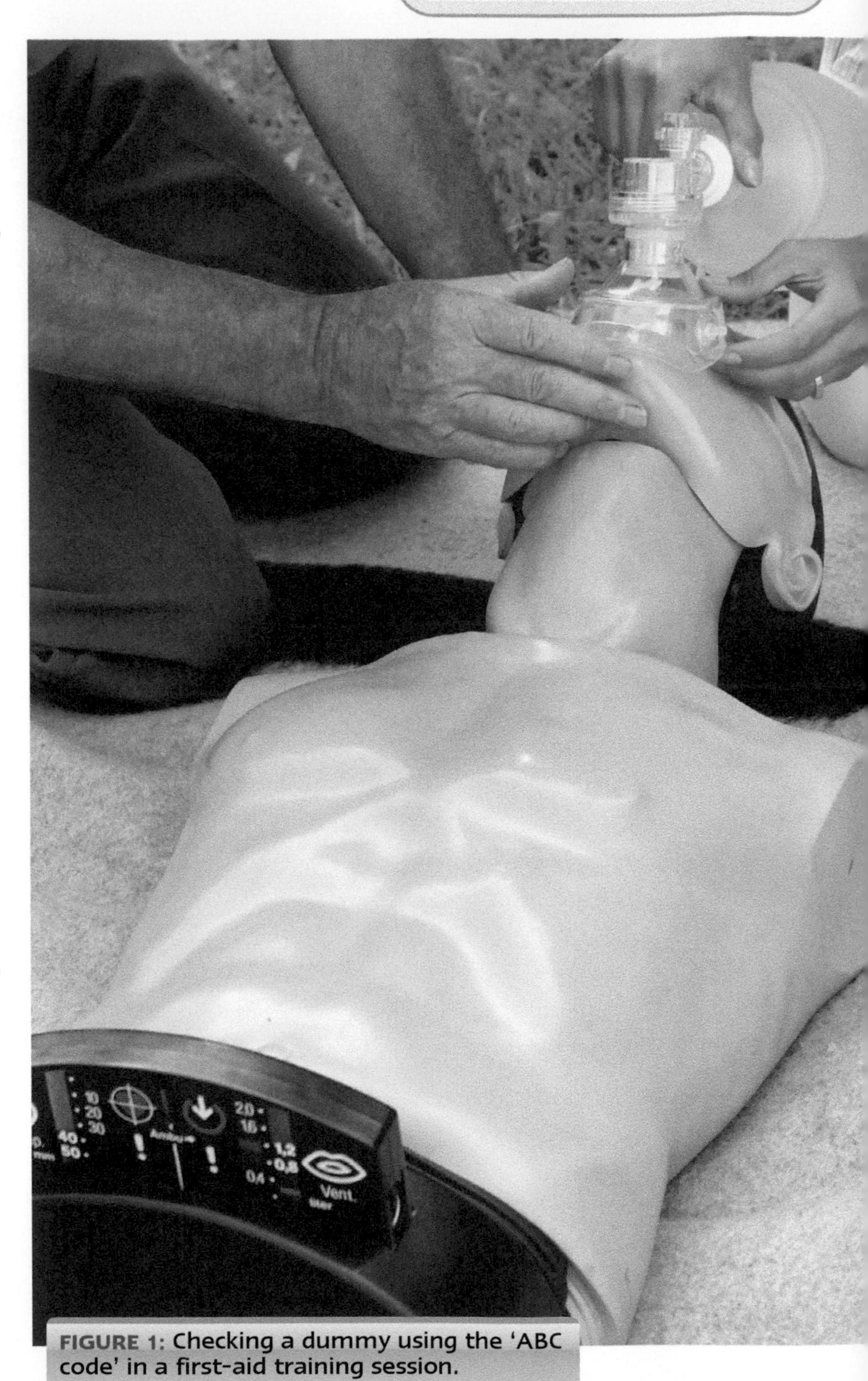

FIGURE 1: Checking a dummy using the 'ABC code' in a first-aid training session.

Get with the beat

Indian gurus are able to lower their pulse rate to what we would consider to be dangerously low levels by thought alone. Harry Houdini the famous escapologist was also able to reduce his pulse when he was carrying out some of his tricks.

You are going to investigate your pulse rate at rest and straight after exercise. You are going to compare the rates of the students in your class.

Your teacher will provide you with the apparatus that you may need for your investigation.

Method:

1. Measure your pulse rate at rest. Make a note of the measurement.
2. As a whole class, collect together all students' pulse rates in a frequency table. Record girls' rates in red and boys' rates in blue.
3. Now draw a frequency graph of the class results.
4. Carry out three different levels of exercise as directed by your teacher. Straight after each period of exercise measure your own pulse rate and make a note of it. (Make sure you are carrying out a fair test by allowing your pulse rate to fall to its 'at rest' level between each exercise period.)
5. Construct a suitable table to record your results in.
6. Now draw a line graph of your results and describe the pattern it shows.

Exam Tip!

When drawing graphs, remember to label your axes (including units if appropriate).

Questions

1. What is the range of resting pulse rates in your class?
2. What is the most common resting pulse rate in your class?
3. Is there a difference in the ranges of resting pulse rates for boys and for girls in your class?
4. Do you think that you have collected enough results to make a conclusion?
5. Explain why a person's pulse rate returns to normal once they start to rest after exercise.

6. Design an experiment to investigate if the pulse rate of teachers is higher in the morning when they get up for work, than in the evening when they get home.

How do you know if you are fit?

BIG IDEAS

You are learning to:

- Describe how you can judge the fitness of a person
- Explain the difference between health and fitness
- Design training programs for certain sports

Training 

If a person turned up to run the London Marathon for charity without having done any **training** beforehand they would be foolish. To do well in a sport a person must train. Each sport needs a different type of **training programme** – hitting a tennis ball each day would not particularly help a person who wanted to run in a marathon.

1. What activity would you do to train for the London Marathon?
2. Suggest what might happen if a person tried to run a marathon without training.

Did You Know...?

Paula Radcliffe ran 150 miles a week to train for the 2002 London Marathon.

FIGURE 1: Runners in a marathon. What sort of weather conditions do you think it would be most comfortable to run in?

Health check

Before a person embarks on a training programme it is important that they have a general **health check** carried out by their doctor. The doctor checks the person's weight, diet, and finds out about previous injuries and any illnesses. All top sports people have **dietitians** who advise them on their diet.

 Make a list of the foods that a person should avoid if they are trying to lose weight.

Fitness

A person can be in good physical and mental condition but might not be **fit** to run a marathon.

The effects of training are:

- increased heart rate and breathing rate, which lead to
- improved circulation
- build up of muscles used for movement
- increase in strength of heart muscle
- increase in strength of breathing muscles.

Fitness is the ability of a person's body to cope with the physical demands placed on it. The better it copes, the fitter the person is.

4. What happens to a person's pulse rate if their heart muscle becomes stronger?
5. What does the blood carry to the muscles?

Measuring fitness

Fitness can be measured by measuring how long it takes for a person's pulse rate to return to normal (its resting rate) after hard exercise. This is called the **recovery time**.

The following results were obtained for Year 9 girls running a 400 metre race.

Name	Resting pulse rate (beats per minute)	Pulse rate after a 400 m run (beats per minute)	Recovery time (minutes)
Mo	66	112	3
Jade	78	128	5
Toni	73	114	3

6 Whose pulse rate increased the most after exercise?

7 Give **one** reason why Toni might be the fittest and **one** reason why Mo might be the fittest.

Progressive training

In order for athletes to improve their performance they have specific training programmes planned for them. The programmes increase in intensity: they are **progressive**.

For endurance events, like long distance running and swimming, the person needs to build up their stamina. This can involve running many miles a week and many repetitions of activities, using light weights. This is mainly aerobic training. For events involving power, a person needs to build up their muscles using heavy weights.

FIGURE 2: What sport do you think this man is in training for?

8 Develop a training programme for a rower in the Olympics. Explain what your programme is trying to achieve.

9 Explain why the pulse rate recovers slowly after exercise.

10 Marathon runners will take regular drinks of fluid containing water, salts and glucose during the run. Explain why they drink these drinks.

How Science Works

Olympic gold medallist rower James Cracknell said, 'Achieving peak physical conditions is one thing, but having a belief in it is another. Mental toughness is just as important as physical preparedness.' Sports psychologists are often used to prepare athletes for the events.

Practice questions

1 Copy the columns below into your exercise book. Match each organ to the correct role it plays in the body.

Organ:	**Role it plays in the body:**
small intestine	remove carbon dioxide from the body
lungs	pumps blood to the tissues
heart	causes movement
muscle	absorbs digested food into the blood

2 The diagram shows the digestive system.

a What letter, **A** to **D**, labels the stomach?

b In which organ is food absorbed?

c What pushes food through the digestive system?

d What is absorbed back into the body in **C** to make the waste solid?

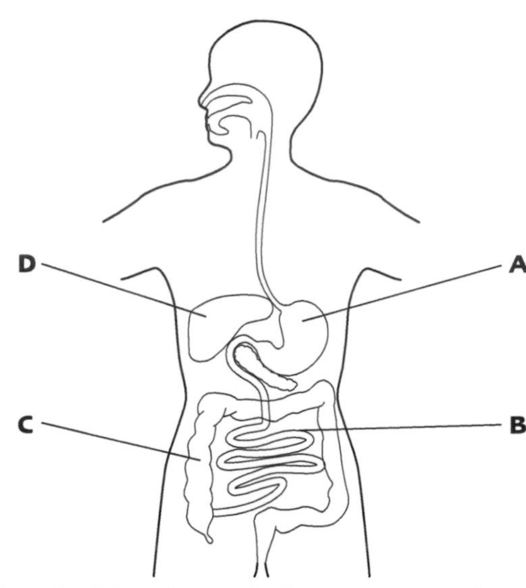

3 Copy the columns below into your exercise book. Match each food type to its correct use in the body.

Food type:	**Main use of food in the body:**
carbohydrates	growth and repair
protein	for energy
fat	insulation and a reserve energy source

4 Luke is studying the human breathing system for a project. He first of all works out the route air takes from the mouth to the blood during breathing.

a Place the words below in the correct order for the path air takes when a person breathes in.

mouth alveoli bronchiole windpipe blood

b Luke finds out that asthma is caused when a person's bronchioles become narrowed. Explain why a person suffering from asthma can become breathless.

c He also finds out that smoke from cigarettes can cause the breakdown of the alveoli, eventually causing a disease called emphysema. Explain why a symptom of this disease is breathlessness.

5 Stefan looked up the composition of the air that he breathes on the Internet. He produced the table shown below.

Gas	Inhaled	Exhaled
Nitrogen	79.00	79.00
Oxygen	20.93	16.10
Carbon dioxide	0.03	4.50

Stefan tried to explain the results. He first looked for patterns.

a i Which gas increases in amount between inhaled and exhaled air?

ii Explain why.

b i Which gas reduces in amount between inhaled and exhaled air?

ii Explain why.

c Explain the difference between breathing and respiration.

6 Amir and Claude were investigating the effect of enzymes on starch.

They knew the enzyme breaks the bonds between the starch to form glucose.

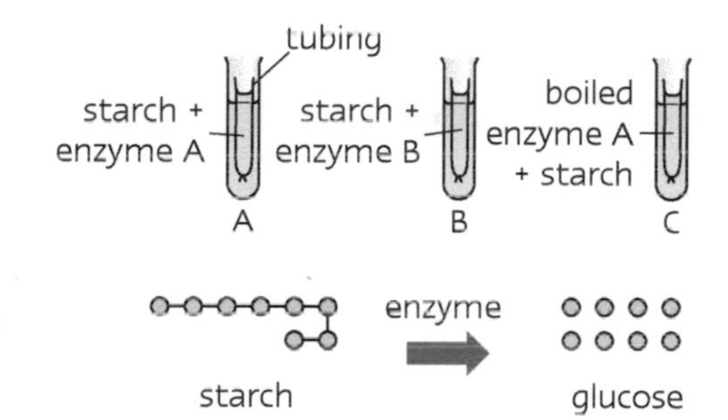

a The experiment was carried out at 35 °C and the solutions left for 20 minutes. Explain why.

b Glucose was found in the liquid in tube A but not in B or C.

i Explain why glucose is found in tube A.

ii Explain why no glucose is found in tubes B and C.

7 Jamal listened to the start of an international rugby match on the TV. The commentator said the players were psyching themselves up to get the adrenalin flowing. He decided to look up what adrenalin did later.

He found out that adrenalin is produced by the body in times of excitement and stress. Its effect is to:

increases the heart rate

increases the pulse rate

increases the amount of glucose in the blood

Use your knowledge of the circulatory and respiratory systems to explain why athletes would perform better when they produce adrenalin.

Learning Checklist

4

- ☆ I can name two types of protein. page 8
- ☆ I can name two types of carbohydrate. page 8
- ☆ I can name two types of fat. page 8
- ☆ I can explain why we need to eat fruit. page 10
- ☆ I know that eating too much salt is bad for the health. page 11
- ☆ I can measure my pulse. page 23

5

- ☆ I know that we need protein for growth, carbohydrate for energy and fats for protection and energy. page 8
- ☆ I know that energy is used for growth, warmth and movement. page 14
- ☆ I know the route air takes when it enters the lungs. page 18
- ☆ I know the parts of the heart. page 20
- ☆ I know some of the causes of heart disease. page 21
- ☆ I know ways of assessing my fitness. page 25

6

- ☆ I can explain why salt is bad for your health. page 11
- ☆ I know the word equation for respiration. page 15
- ☆ I know the role of cilia and mucus in cleaning the lungs. page 18
- ☆ I know where oxygen enters the blood and is transported to the cells. page 19

- ☆ I know the difference between breathing and respiration. page 19
- ☆ I can explain why the pulse rate increases with exercise. page 23

- ☆ I can explain the effect of a lack of enzymes on a person's health. page 13
- ☆ I can explain how the working of a pump and the heart are similar. page 21

Topic Quiz

1 Here is a list of foods.

meat **bread** **butter** **milk**
fish **nuts** **sweets** **potato**

Which of the foods are rich in:

a carbohydrates
b protein
c fat?

2 What happens to a person's pulse rate when they take exercise?

3 Name **two** foods rich in iron.

4 What is the role of the heart?

5 What are the units of energy?

6 How are the lungs self-cleaning?

7 If the artery leading to the heart becomes blocked by a blood clot why does the person suffer a heart attack?

True or False?

If a statement is false then rewrite it so it is correct.

1 Starch and sugar are carbohydrates.

2 Calcium is needed to make red blood cells.

3 The lungs are like plastic bags.

4 The heart pumps blood to the lungs.

5 A heart attack is caused when the blood vessel to the heart is blocked so the heart cannot get rid of carbon dioxide.

6 A stroke affects the heart.

7 Red blood cells carry oxygen.

8 Digestion is the break down of large food molecules to smaller molecules.

9 Digestion is carried out by enzymes.

10 Respiration involves using carbon dioxide to break down food to make energy.

11 The alveoli are lined with cilia and mucus.

Literacy Activity

The inhaling of dust causes lung disease. In miners dust on the lungs or pneumoconiosis is caused by inhaling coal dust produced by machinery and digging. During a lifetime a miner may inhale up to 1000 g of dust.

Asbestosis is caused by inhaling asbestos fibres. The disposal of asbestos requires special health and safety procedures. Silicosis results from inhaling sand or quartz dust. Symptoms of dust on the lungs are not always shown, although early symptoms are shortness of breath followed by a cough and possible chest pains. Later on lung scarring occurs as the alveoli are damaged.

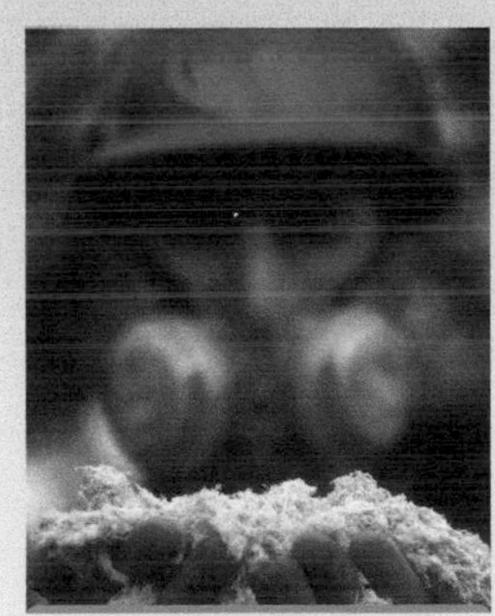

Why is the person holding this asbestos wearing protective clothing and special breathing apparatus?

1 Which industries may cause dust on the lungs?

2 How are the lungs able to self-clean?

3 What causes a person to have:
a breathlessness **b** a cough?

4 Find out if there is any treatment for these diseases.

Microbes and disease

BIG IDEAS

By the end of this unit you will be able to describe types of microorganisms, their uses and the hazards they represent. You'll be able to explain the role of vaccinations in fighting disease and the implications this has for society.

Malaria

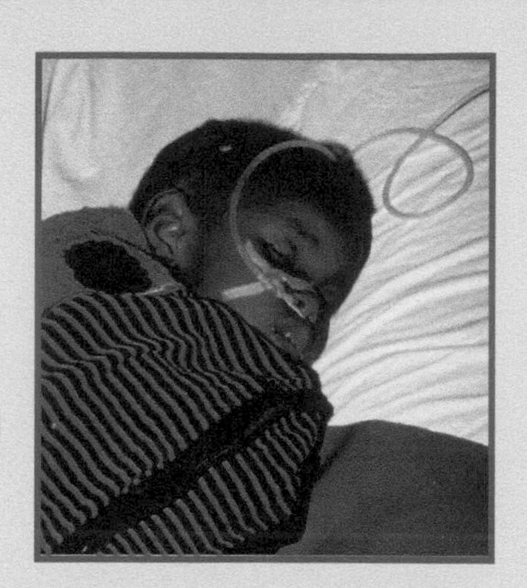

The mosquito carries the parasite and when it sucks a person's blood it injects the parasite into the person's body. The parasite invades red blood cells.

The symptoms are hot and cold fevers, sweating and shivering.

Athlete's foot

The fungus is spread through damp conditions or by sharing towels with an infected person. The symptoms are peeling of the skin and the occurrence of open cuts.

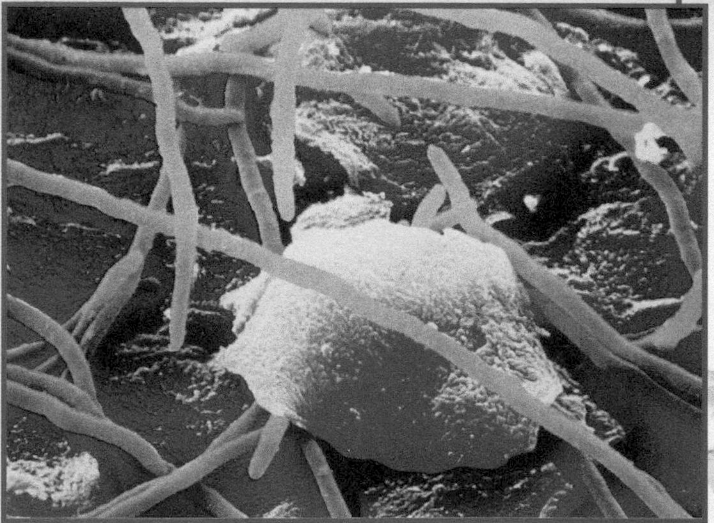

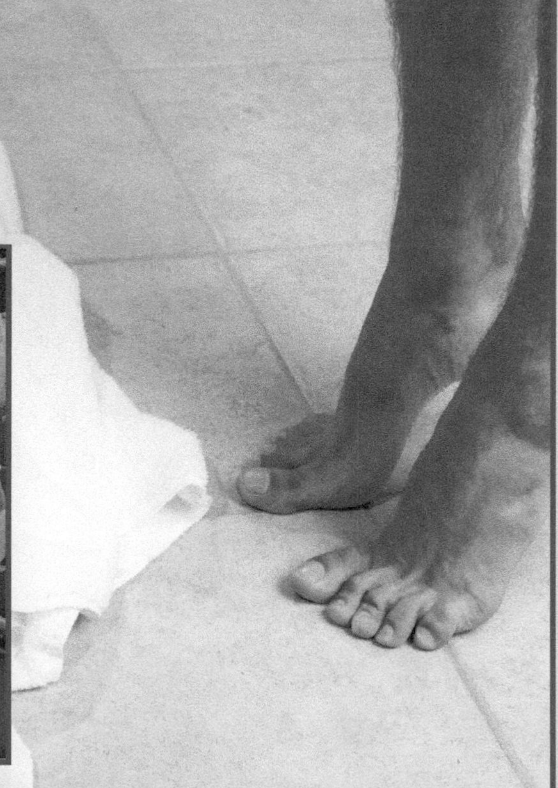

Microbes cause disease, which can be spread. They are called infectious diseases. Some diseases such as diabetes, asthma, hay fever and cancer cannot be spread from one person to another. They are called non-infectious diseases.

In the developing world, infectious diseases and starvation are the major causes of death. They cause a high death rate in infants and also a low life expectancy. There is little food to give people strength to fight infectious diseases. The lack of food can occur because of pests eating the crops, droughts drying up the land and other catastrophes destroying the crops or washing away the soil. When natural catastrophes occur, for example earthquakes and tsunamis, there is also a real risk of disease spreading. Sewage starts mixing with fresh water, increasing the likelihood of the spread of cholera, typhoid and dysentery. Dead bodies start decaying in the warm conditions attracting rats and flies and the stench becomes unbearable. Medical support will be urgently needed, but cannot always be organised quickly enough.

In developing countries there may be little cleanliness (sanitation) and few medical facilities, often rubbish is dumped in the street and sewage may go straight into the rivers. These conditions are ideal for rats and flies to breed, which help to spread the microbes that cause disease. Providing support for people is not helped by the difficulty in distributing supplies due to the lack of good roads.

In tropical countries there are many animals that carry disease, as the conditions are warm and moist. The anopheline mosquito spreads malaria and the tsetse fly carries a disease called sleeping sickness. Many people in Africa and Asia appear to be tired, thin and lethargic.

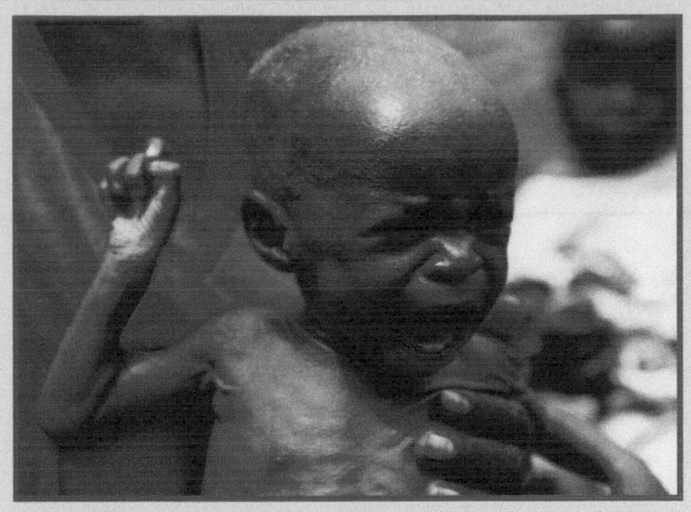

These people are suffering from disease and lack of food. Diseases like HIV are spreading uncontrolled due to lack of health education and unsafe sex between men and women. Children are born with a future of living with HIV as the disease is passed from the mother to the foetus. The sight of a child slowly dying can be heart wrenching.

In developed countries the death rate amongst infants is far lower and the life expectancy far longer.

What do you know?

1. Give the name of **two** diseases that are spread from one person to another person.
2. Name **two** diseases that cannot be spread.
3. Give **two** ways diseases may be spread.
4. What problems are caused by natural disasters?
5. Why is life expectancy low in developing countries?
6. Which organisations provide aid to developing countries?
7. Why is the death rate lower and life expectancy longer in developed countries?
8. What is the difference in role between a nurse and a doctor?
9. In Britain, local hospitals have been closed and larger city hospitals have been built. Give **two** reasons why this has occurred and **two** problems it may have caused.

History of disease

BIG IDEAS

You are learning to:
- Explain the causes of disease
- Explain the implications of Pasteur's work (HSW)
- Interpret the work of scientists on microbes

5

FIGURE 1: What do maggots turn into?

Spontaneous life

Meat goes rotten and maggots are found on the meat. It seems that the meat has created life. This was the theory of **spontaneous** generation of life.

Disease is spread when there are dead bodies around. Could it be the smell given off by diseased and rotting corpses spreads disease? This was another of the beliefs for the spread of disease.

In 1854, John Snow (a British physician) traced an outbreak of cholera in London to sewage-**contaminated** water and not to the dead bodies found. He was able to trace all the cases to the water taken from a certain well. He showed that people living near each other only caught the disease if they used that well.

FIGURE 2: John Snow.

1. Why are maggots found on rotten meat?
2. What could have been done to make the contaminated water safe to drink?
3. How did Dr Snow's work show that disease was not spread by the smell of dead bodies?

Pasteur's work

FIGURE 3: Louis Pasteur.

Pasteur was one of the great French scientists. He did not believe life was created from rotten meat. In 1860, he set up flasks containing meat broth and left them to see if they rotted. The apparatus he used is shown in Figure 4. He put a broth solution in the flask and boiled it until all life was destroyed. He left the liquid to stand for many months. The liquid remained clear. He looked down his microscope and saw no life and no sign of rotting. If he broke the neck of the flask, the liquid went cloudy.

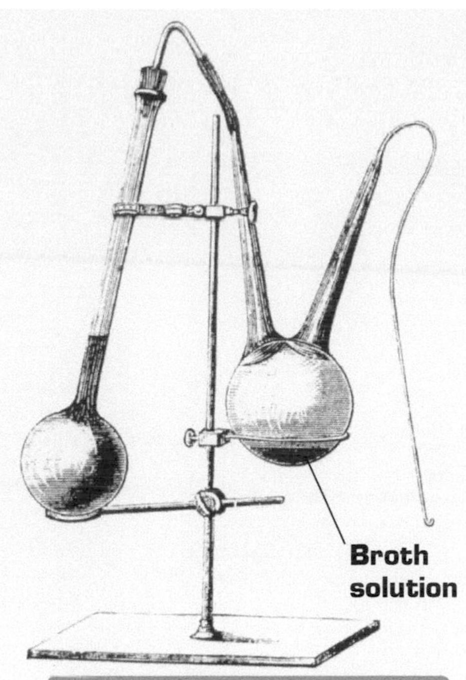

FIGURE 4: Pasteur's apparatus.

4 Why did the meat broth not rot?

5 Why did the broth go cloudy when he broke the neck of the flask?

6 How did his work disprove the idea that rotten meat makes life?

How Science Works

Koch was a very brave scientist because he tested out an early vaccine for tuberculosis on himself. He became very ill and nearly died. Today, this would be regarded as unethical. HSW

A German called Robert Koch in the 1800s showed that anthrax was due to long filaments that formed spores. Anthrax was a frightening bacterial disease in which the animal's temperature would rise then the animal would tremble, gasp for breath and usually die. He then showed that the recurrence of anthrax in sheep grazing in pastures that had not been used for many years was due to these spores. This information led to Koch proposing an important principle for science.

If a microbe were the cause of a disease then it would be found in all organisms having that disease.

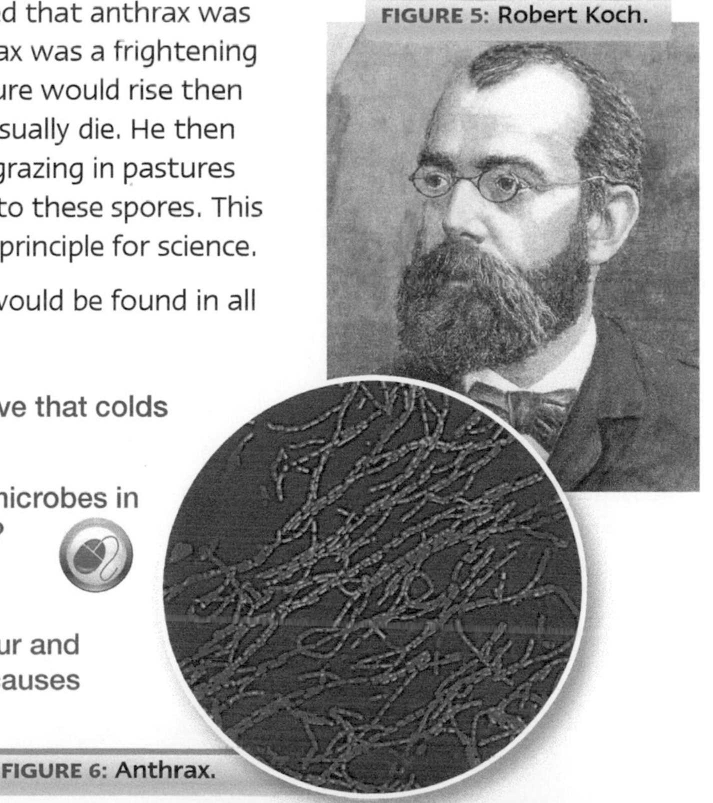
FIGURE 5: Robert Koch.

FIGURE 6: Anthrax.

7 Using Koch's principle, how would you prove that colds are caused by a different microbe to flu?

8 How might Dr Snow have found out if the microbes in the contaminated water did cause cholera?

9 Explain how the work of John Snow, Pasteur and Koch led to a better understanding of the causes of disease.

The infection cycle

BIG IDEAS

You are learning to:
- Explain the difference between infections and symptoms
- Explain how diseases can be spread
- Explain the link between allergies and infection

How do we catch a disease?

Microbes are microscopic organisms for example bacteria, viruses and fungi. They can be spread in several ways:

Means of spread	Details	Name of a disease
air	droplets in the air	flu, cold
water	contaminated by sewage	cholera, typhoid
food	microbes infect food	food poisoning
contact	touch	smallpox, leprosy
animals	mosquitoes	yellow fever
	tsetse fly	sleeping sickness
	rabid dogs	rabies

1. How do you catch a cold?
2. Why does boiling water stop the spread of typhoid?
3. Which **two** diseases are likely to cause problems when earthquakes or hurricanes hit an area?
4. Why does covering your mouth when you sneeze help prevent colds from spreading?
5. For each way in which microbes are spread, suggest a way of controlling the disease.

Did You Know...?

Head lice spread the disease Typhus. To reduce its spread in World War I, soldiers slept head to toe in the trenches.

So what are bacteria and viruses?

A bacterium is a single celled organism that is a few micrometres long. They live in the human body in huge numbers, mainly on the skin and in the digestive system.

A virus is smaller and is not a complete cell. Unlike bacteria they cannot be treated with antibiotics but are often eliminated by the body's immune system, which can be developed by vaccines.

... *allergy* ... *incubation period* ... *infection*

Infection 

Not all microbes cause disease, only the **pathogens** (disease causing microbes).

The first stage of **infection** is the entry of the pathogens into the body, past the external defences. The pathogens enter the cells and multiply, causing damage to the cells and releasing poisons (**toxins**). At this stage the body's immune system is stimulated to begin working and **symptoms** start to be shown. The period between when the person becomes infected and when the symptoms of the disease start to show can be the **incubation period**. A common symptom is fever. Fever can be caused by the white blood cells releasing chemicals when they act on pathogens.

6. Suggest why it is important to stay at home when you have flu.
7. What is the difference between becoming infected and showing the symptoms?
8. Explain why it is possible for an infected person who has no symptons to spread a disease.

Did You Know...?

Viruses such as these are the smallest microbes with a size of 0.02–0.025 micrometres.

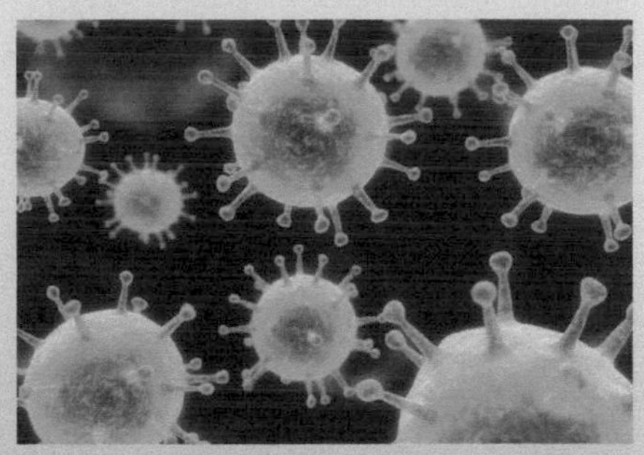

Bacteria are 2–10 micrometres in size.
A micrometre is 1/10 000 000 metres.
About 1000 bacteria would stretch end to end across a pin top.

Allergies

An **allergy** is caused when our immune system overreacts or reacts to the wrong substance like dust, pollen, chlorine or nuts. The symptoms may be itching, wheezing, sneezing and watering eyes. The symptoms occur because the person has become sensitised to the foreign material. In the case of allergy to bee or wasp stings or nuts the response can result in death due to shock to the circulatory system or the inability to breathe. These are anaphylactic reactions and a person who is likely to go into anaphylactic shock carries a 'pen' for injecting chemicals to prevent death.

9. Explain how hay fever symptons are caused using your knowledge of the infection.
10. Explain why people sneeze when they have hay fever or a common cold.

FIGURE 1: Allergies to nuts are becoming more common.

FIGURE 2: Why are wasp stings fatal for some people?

FIGURE 3: At what time of year is hay fever most common?

... pathogens ... symptoms ... toxins

Preventing disease

BIG IDEAS

You are learning to:

- Describe the mechanisms used by the body to prevent disease
- Explain why you rarely catch a disease twice
- Explain the importance of white blood cells

A barrier to bacteria

Microbes are all around you, but you do not keep suffering from diseases. A person has ways of stopping microbes getting into their body. The first line of defence is the skin, where the outer cells act as a barrier.

The skin can be broken when you cut or graze yourself. Microbes can now enter the blood. To stop the loss of blood and prevent microbes entering, the blood will clot then dry and go hard. This forms a **scab**.

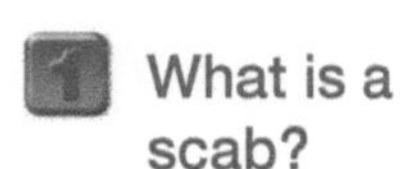

1. What is a scab?

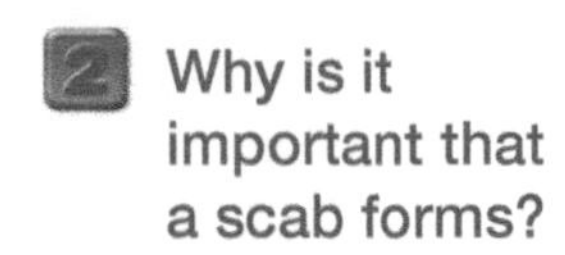

2. Why is it important that a scab forms?

Did You Know...?

Haemophilia is a disease where a person's blood does not clot properly. They have to go to hospital if they have a deep cut to stop blood flow.

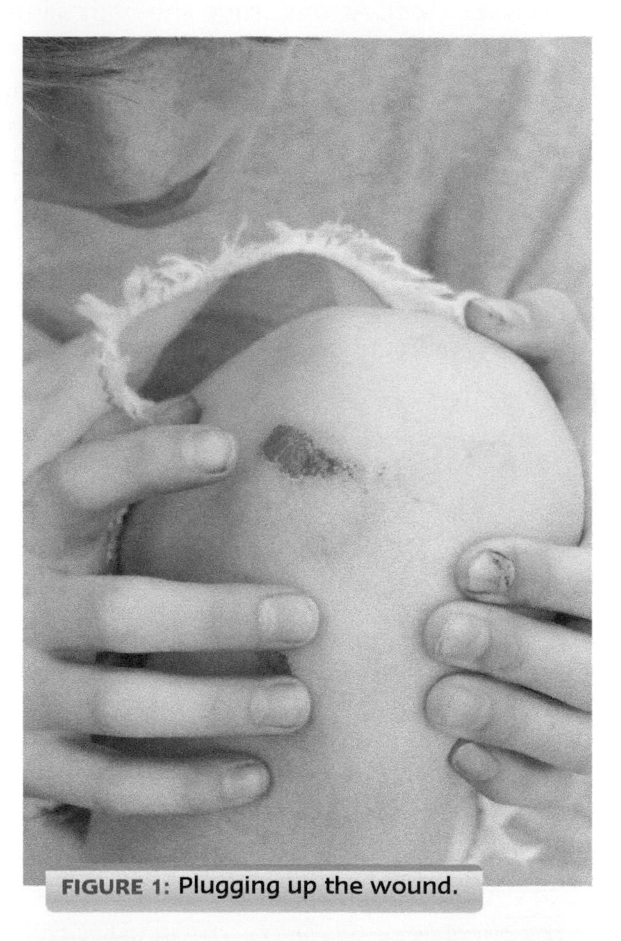

FIGURE 1: Plugging up the wound.

First line of defence

If microbes enter the lungs they are trapped in the mucus lining and then swept out by cilia. If they enter the stomach the acid conditions kill them. Even the eye produces tears that kill bacteria.

The body's next line of defence is the **white blood cells**. These are the body's army, they are brought into action when the first line of defence is breached.

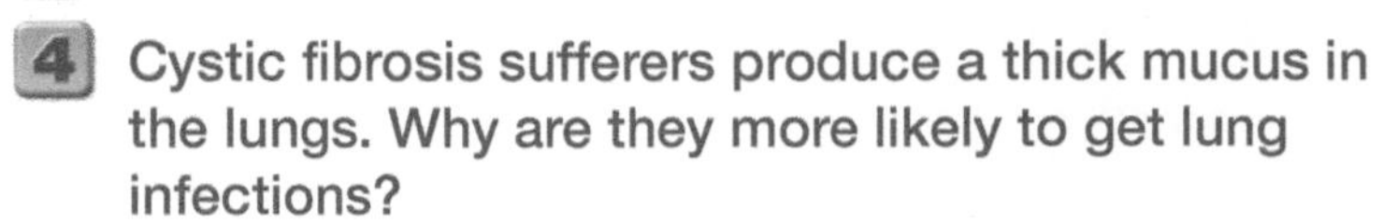

3. Give **two** ways the body stops microbes entering.
4. Cystic fibrosis sufferers produce a thick mucus in the lungs. Why are they more likely to get lung infections?

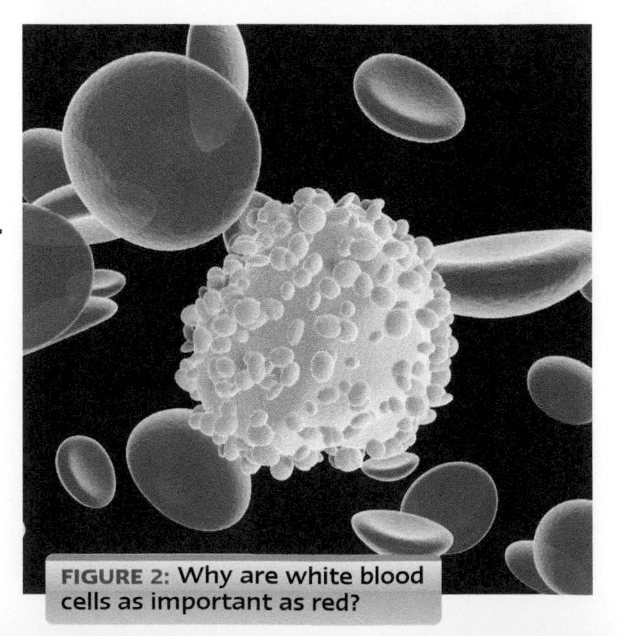

FIGURE 2: Why are white blood cells as important as red?

Mobilising the army

The army of different types of white blood cells have several ways of fighting infection once their defences are called on. One type surrounds and digests the microbes.

Another type produce chemicals called **antibodies**, which attach to the microbes and destroy them. A third type produce chemicals called **antitoxins**, which neutralise poisons (**toxins**) produced by the microbes. The white blood cells gather in the area of infection (where the microbes entered) and the battle commences. More blood is directed to the area to allow more white blood cells to get to the microbes. These mechanisms of defence are called the body's **immune system**.

5 What is the difference between antibodies and antitoxins?

6 Why will an area of infection go red and feel hot?

Exam Tip!

Antibodies are chemicals produced by white blood cells to kill microbes.
Antibiotics are chemicals produced by microbes to kill other microbes.

How Science Works

Antibodies are able to recognise specific proteins. Monoclonal antibodies are used in pregnancy testing where they identify a protein in the urine of a pregnant woman. HSW

Stop it happening again 

When a person has a disease the body prepares defences in case the disease attacks the person again. The white blood cells produce a special type of white blood cell called a memory cell. These enable antibodies to be produced very quickly on the second infection. This usually means you rarely get a disease twice because the antibodies are produced in large amounts far quicker.

7 The diagrams show how white blood cells fight microbes. Give the strengths and weaknesses of these diagrams in how they show white blood cells at work.

8 Explain why the first time you get a disease, it is far worse than the second time.

9 Use the idea of how the body prevents disease to explain why a person is resistant to a disease when they have never had it and the same disease could kill someone the first time they have it.

FIGURE 3: The antibody army attacks.

Bacteria enter cut

Antibodies are produced

Antibodies fight microbes

Sexually transmitted diseases

BIG IDEAS

You are learning to:

- Describe the different types of STD
- Explain the difference between AIDS and HIV
- Explain the problems caused by unsafe sex

4

Lice

One way of spreading disease by touch is through sexual intercourse. There are at least 25 diseases that can be spread during sex, including lice. Lice found in pubic hair can spread to a sexual partner. The lice will cause itching.

1. How will washing clothing and bedding reduce the spread of lice?
2. What does sexually transmitted diseases mean?

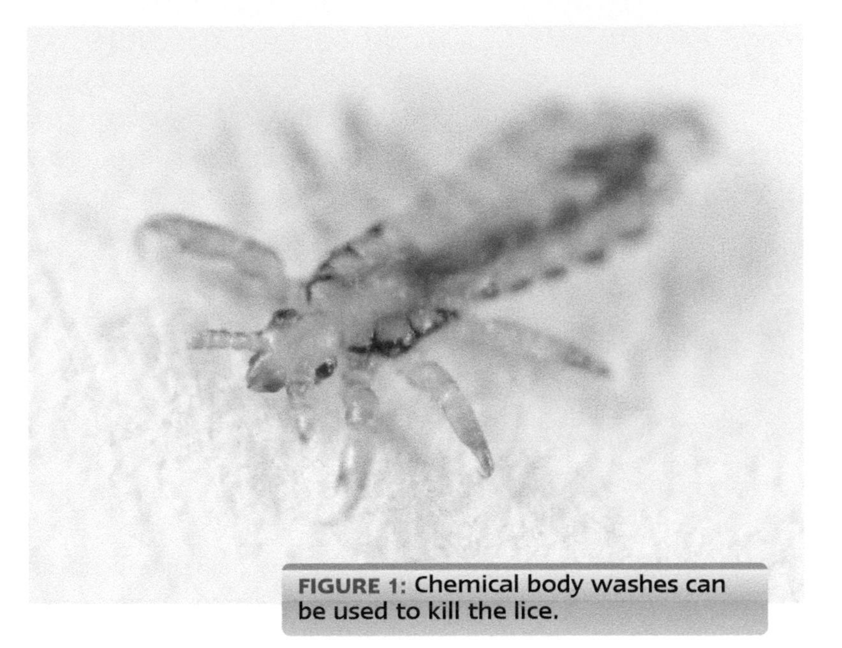

FIGURE 1: Chemical body washes can be used to kill the lice.

5

STD

There are references to most sexually transmitted diseases in history books e.g. the pox, which is now called **syphilis**, so they are not new diseases. The table below is a brief summary of the more common **STDs**.

Disease	Microbe causing the disease	Symptoms	Cure
Syphilis	Bacteria	Primary stage is usually shown by painless sores that last a few weeks. The secondary stage is a rash which again disappears. Finally damage occurs to the brain, eyes and heart with death occurring.	Antibiotics

Disease	Microbe causing the disease	Symptoms	Cure
Gonorrhoea	Bacteria	There may be no symptoms, but in certain cases pain occurs when urinating. If untreated infertility can occur.	Antibiotics
Thrush	Fungus	Itching, swelling and soreness.	Antifungal creams and washing of infected areas.
Genital Herpes	Virus	Itching, tingling sensations.	
Chlamydia	Bacteria	Often no signs are seen. It can cause infertility in women.	

3 How are the two diseases spread by bacteria treated?

4 Why is it difficult to tell if you have Gonorrhoea?

5 What are the risks of having mutiple sexual partners?

HIV and AIDS

The spread of **HIV** is fairly recent compared to other sexually transmitted diseases. HIV is the name of the virus that causes **AIDS**. The virus can be spread by the mixing of sexual fluids or blood. Infected blood used in transfusions was also responsible for the spread of HIV until the blood was tested before it was used. Many drugs have been used to slow the development of AIDS after infection with HIV, but not cure it.

6 What is the difference between HIV and AIDS?

7 Why might a drug user who shares needles get HIV?

Onset of AIDS

HIV stops the immune system from working. This means other diseases can then attack the body. Skin infections, weight loss, tuberculosis and rare forms of cancer result in the symptoms shown by AIDS. The main ways of preventing HIV spreading are to have one partner at once and practise safe sex, which means a condom is used to prevent the mixing of fluids.

Did You Know...?

AIDS first appeared in the news in 1981. AIDS now kills more people between the ages of 15–49 than any other disease in the world.

8 At what stage do the symptoms of AIDS occur?

9 Both HIV and syphilis are spread by sexual intercourse. Suggest why HIV has caused more problems than syphilis.

Congo's Ebola town

The town in the centre of an outbreak of the deadly Ebola virus in the Congo has been sealed off to stop the disease spreading, health officials say. Nine people have died in the northwestern town of Etoumbi. The World Health Organisation director said food would be delivered to the town. He indicated there was no known cure for Ebola also known as haemorrhagic fever.

'The Head of the district has placed the town in quarantine, nothing goes in or out any more' said one of the doctors.

The outbreak appears to have begun when some villagers came across the body of a chimpanzee and took it back to the town. The meat is a delicacy, even though villagers have been warned not to touch dead bodies.

Ebola is highly contagious and is spread through body fluids such as sweat and blood.

Our science correspondent reports that Ebola is regarded as one of the worst diseases in the world. It is usually characterised by a sudden onset of fever, intense weakness, muscle pain, headache and sore throat, followed by vomiting, diarrhoea, a rash, liver and kidney failure and massive internal and external bleeding. It kills about 90% of those infected.

So which is the worst disease?

Ebola

Malaria

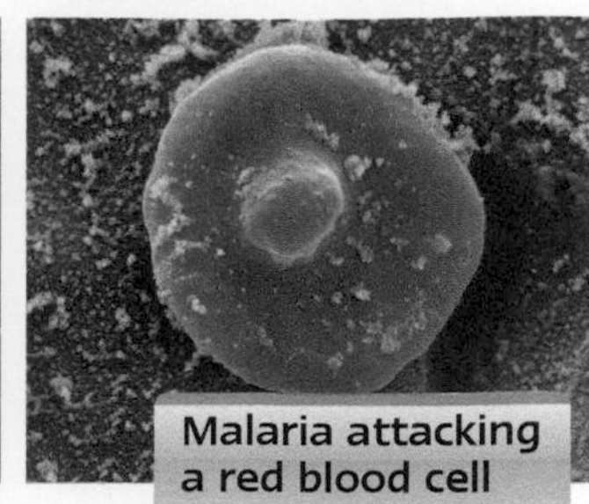

Malaria attacking a red blood cell

HIV

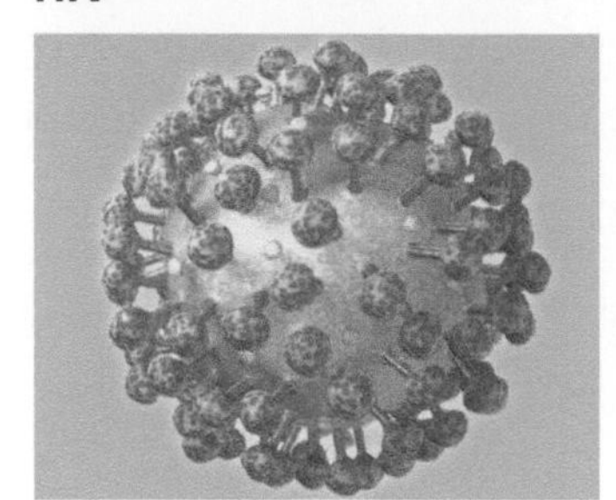

Flu

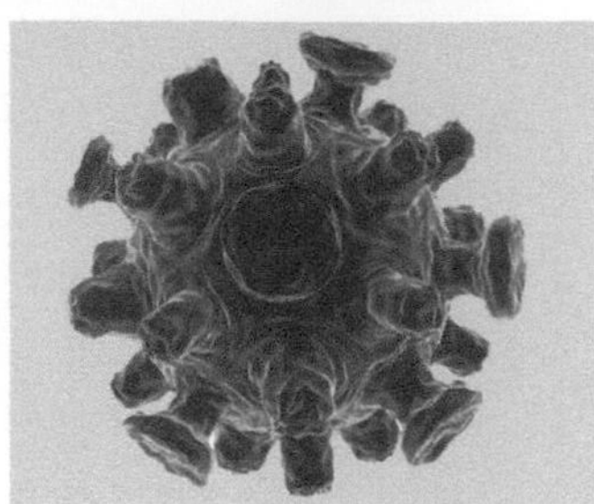

Smallpox

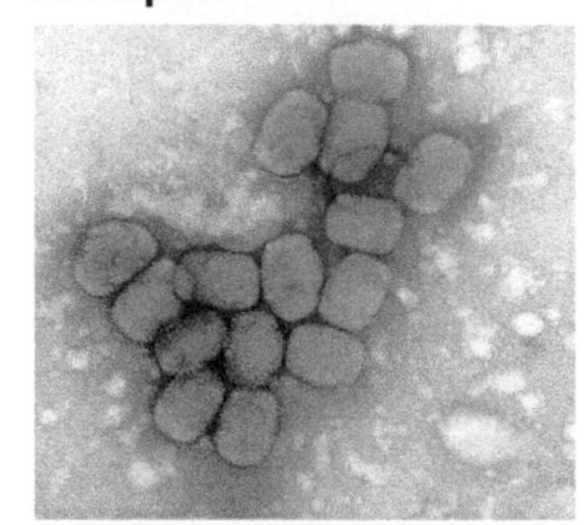

Anthrax

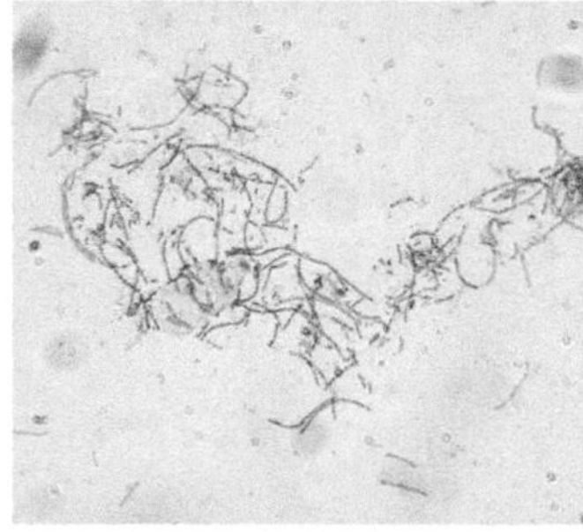

Letters

From Dr Stephen Taylor.

'Although Ebola kills most people who catch it, it is not able to spread rapidly and kills few people every year. Malaria and HIV must be regarded as far more dangerous diseases as they kill over a million people a year. Most of these die in Africa. HIV has at present no cure.'

From Dr Paula Bent.

'If we look back in History there were many epidemics that spread rapidly killing many people. The Black Death or the Plague spread throughout Europe killing hundreds of thousands of people. Anthrax was a frightening disease in the nineteenth century as if you caught it you would usually die. One in three people who caught smallpox died in the seventeenth century. There have always been horrible diseases. I am sure there will be other similar diseases in the future.'

Dr Pete Hatcher.

'How can you judge the worst disease? Is it the death rate of those infected? Is it how easily it spreads? Is it whether it can be cured? Is it on the symptoms it causes? I can make a very strong case for the common cold or even flu being the worst disease. They both spread rapidly, are difficult to control and flu injections are needed to prevent deaths of the elderly.'

Bob Light, science correspondent.

'Many of the most dangerous diseases in the world are kept at research laboratories under high security. The diseases are given ratings to indicate the level of precautions that must be taken when they are studied. This will mean scientists must dress in protective clothing and go through various sealed rooms to reach the disease to be studied.'

FIGURE 1: Why is this scientist wearing a breathing mask and protective clothing?

Assess Yourself

1. Which diseases are the biggest killers of people in the world?
2. Explain how the following diseases are spread:
   Ebola
   Flu
   Malaria
   Plague
3. Why do diseases spread more quickly nowadays from country to country?
4. When flu spreads it causes epidemics. What is an epidemic?
5. How does the body stop microbes from entering?
6. How does the immune system act on microbes that have started to attack cells?
7. What precautions are taken when keeping dangerous diseases in laboratories?
8. What do you think are the key points in judging how dangerous a disease is?
9. Use the key points to decide which disease is the worst. Explain the reason for your choice.

ICT Activity

Prepare a Power Point on a different disease to show its effect on humans.

Level Booster

8 Your answers show you can use the key points to evaluate the dangers of each disease. You are then able to justify clearly which is the worst disease, having considered several other diseases.

7 Your answers show you can suggest key points that can be used to judge diseases. You can then give reasons for choosing which disease is the worst.

6 Your answers show you understand how the immune system works.

5 Your answers show you understand why diseases spread more rapidly nowadays. You can also name two ways the body prevents microbes from entering.

4 Your answers show you understand how diseases are spread.

Biological warfare

BIG IDEAS

You are learning to:

- Describe how biological warfare has been used
- Explain the procedures used to stop dangerous diseases from spreading
- Discuss the ethics of biological warfare 

No water 

The poisoning of wells with dead bodies by the Spartans in the siege of Athens in 404BC and the poisoning of spikes in traps by the Viet Cong in the 1960s in Vietnam are all examples of biological warfare. Biological warfare is the use of microbes (bacteria, viruses and fungi) to kill the enemy.

1. Why were the wells poisoned?
2. Spikes can kill. Why were they tipped with poison?

Did You Know...?

Hannibal in 190BC won naval battles by hurling earthen bottles filled with venomous snakes on to enemy boats. He also used elephants as their size, smell and trumpeting scared the Roman horses.

Arrows of death

In 404BC, Scythian archers from the Black Sea dipped their arrows in a mixture of human blood, dung and the decomposed bodies of venomous snakes. The poisons contained tetanus and gangrene bacteria. The archers could fire 20 arrows in a minute to a distance of 500 metres.

In 1767, during the wars between the French and English in North America, the English general gave the Native Americans supporting the French blankets infected with smallpox. This caused the disease to spread amongst the Indians resulting in a lot of deaths.

3. How is smallpox spread?
4. Explain whether you think the general was right to give the Native Americans blankets.

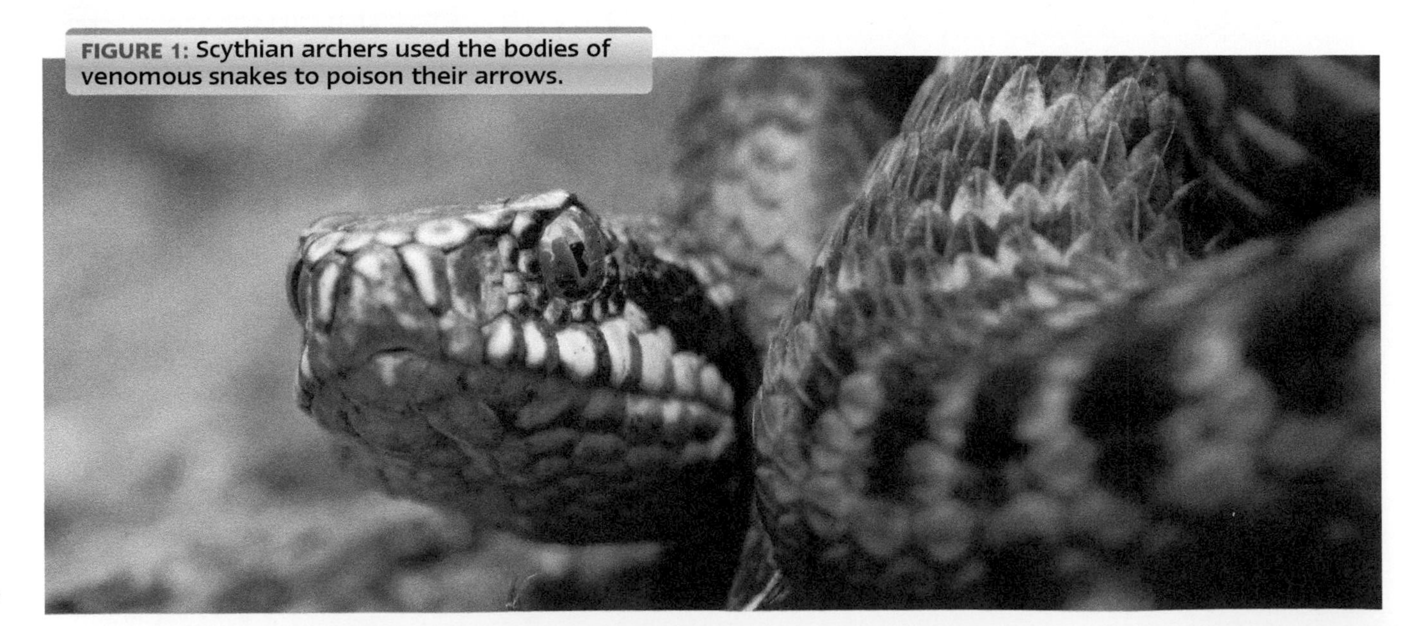

FIGURE 1: Scythian archers used the bodies of venomous snakes to poison their arrows.

Use of anthrax and salmonella

Evidence exists of anthrax being used by the Germans in World War I and in trials by Britain during World War II. In America in 1984 salmonella was sprinkled in a salad bar. This was to infect people and stop them voting in the local election. It was a trial run for a later attack. The trial was successful: with over 750 cases of food poisoning, 45 requiring hospitalisation. The people were caught and found to be members of a minority party in the election.

How Science Works

Hens' eggs are the ideal sterile media to grow viruses in. Viruses need the living conditions in the egg.
Why are the conditions sterile inside an egg? Why are viruses stored in hens' eggs in laboratories?

HSW

 The island in Scotland where the trials on anthrax took place is still uninhabitable because of spores in the soil. Suggest what experiments they could carry out to find out if it is safe to breed cattle on the island.

The deadly mail

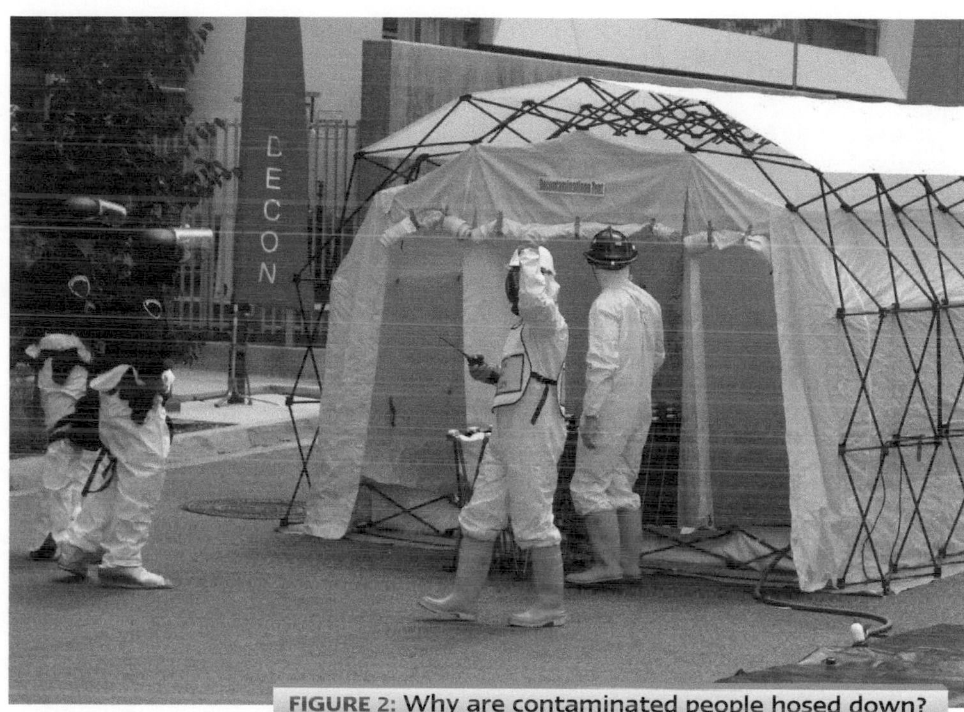

FIGURE 2: Why are contaminated people hosed down?

In 2001 there was great concern about the use of anthrax spores by suspected terrorists. Letters were sent through the post and the spores were released when the letters were opened. Eighteen people were infected and five died from the spores being inhaled.

When there is a suspected outbreak of a dangerous disease it must be reported. If the disease is likely to spread rapidly the area will be **isolated** and **quarantined**, so people cannot get in or out. People may be forcibly kept in the area.

Scientists wear special suits and places are set up where the contaminated people can be washed down with chemicals, these are called **decontamination** areas.

 Explain in detail how the area is quarantined.

 Give a reasoned account to explain whether you think people should be kept in quarantine if there is an outbreak of a deadly disease, even though they may die.

 Is biological warfare a legitimate use of science? 

Vaccination

BIG IDEAS

You are learning to:

- Explain how early vaccines were given
- Interpret information about how vaccines work HSW
- Discuss the problems in explaining ideas to the public HSW

The spread of smallpox

In the 1600s **smallpox** was a killer disease. One in three people who caught smallpox were likely to die from this disease. No cure was known. Smallpox has now been wiped out from Britain.

1. Describe the signs of smallpox shown in the picture.
2. How many children might have died in the 1600s in a class of 30?

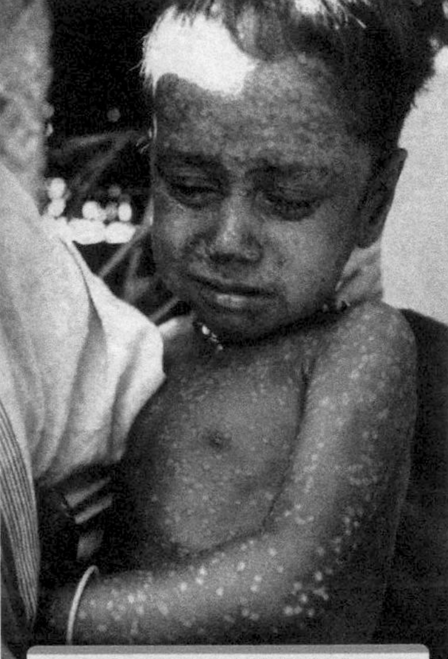

FIGURE 1: A child with smallpox. People still die from smallpox in Africa.

Early treatments

Lady Montagu was travelling in India and saw that children were being given scrapings of smallpox from people who were not seriously affected; this was thought to be a weakened dose. Some of the children caught smallpox and died, others survived and never caught smallpox again. She came back to Britain and tried experiments with smallpox on prisoners to see if they could survive the disease.

3. Why did some of the children die?
4. Why did they use prisoners to test the 'weak dose of smallpox'?
5. Why did they think the dose was 'weak'?

Jenner's work HSW

Edward Jenner was a country doctor. He noted a common observation that milkmaids who caught a disease called cowpox did not catch smallpox. Jenner tested his theory by inoculating a boy called James Phipps with material from a cowpox blister. James became ill but recovered. Jenner then gave James a dose of smallpox by scratching it onto his arm. James did not suffer any illness. This was one of the first cases of the use of a **vaccine**. Jenner strongly believed that cowpox prevented smallpox and he **inoculated** his own son with cowpox.

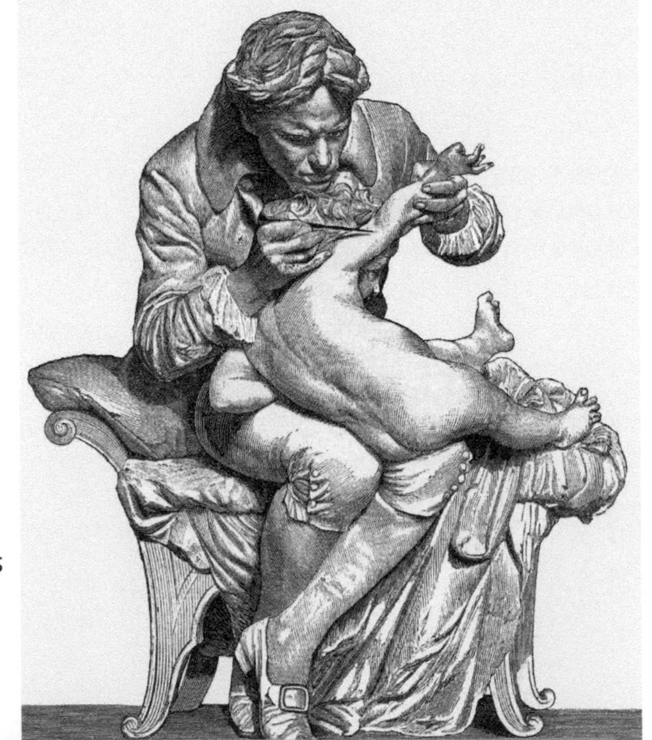

FIGURE 2: Why did Jenner inoculate his own son?

... *immune* ... *inoculated*

6 Explain Jenner's theory about how the vaccine for smallpox worked.

7 Use the model for the immune response to explain how the vaccine worked.

8 Jenner carried out experiments on a boy. The boy could have died.
   a Should experiments be carried out on patients?
   b Suggest why humans were used by Jenner instead of animals.

What the papers said 

Scientists did not consider Jenner's findings seriously at first. After further work testing his idea, Jenner published a report of 23 cases to show that people were **immune** to smallpox when treated with cowpox. The scientists of the time were still very cautious and took a long time to accept the idea. The papers of the time were willing to criticise and published a now famous picture.

FIGURE 3: What impact do you think this cartoon would have had?

9 **a** What was the response of the press to Jenner's discovery?
   b Why do you think they responded in this way?
   c How do you think the press would report his findings today? Write an account of his findings for a named paper.

10 How did Jenner set out to prove his critics were wrong?

Did You Know...?

The last case of a person dying from smallpox in Britain was in 1978 when smallpox escaped at a research laboratory.
Smallpox is almost wiped out throughout the world. It should be gone forever in five years.

What are vaccines?

BIG IDEAS

You are learning to:
- Explain how a vaccine works
- Describe the different types of vaccines
- Evaluate the evidence for the use of vaccines 

A jab of help

When you are young you will be given vaccines for various diseases. The most common ones are measles, mumps, rubella and polio. As a teenager you may also be given a test to see if you are immune to TB, and if you are not, a TB jab will be given.

 What does TB stand for?

Did You Know...?

The early form of vaccinations involved scratching the disease onto a person using a pin. The syringe had not been invented.

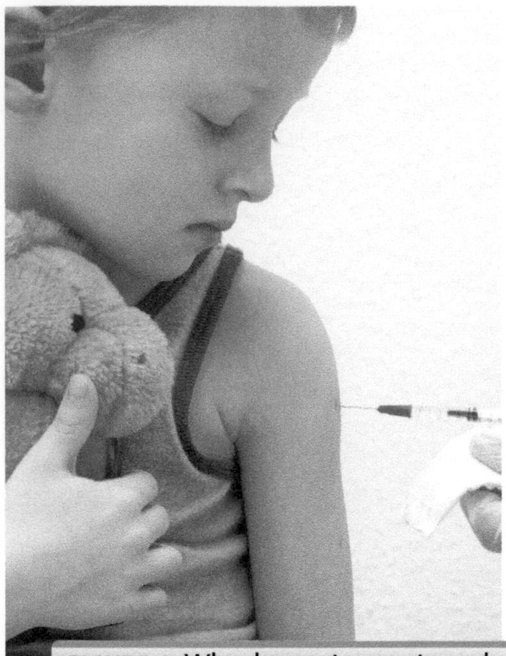

FIGURE 1: Why do most parents make sure their children are vaccinated?

Travelling abroad

Vaccines are also given to people who travel to places where a disease is common. When you go to certain parts of Africa you might have typhoid vaccines, or if you travel to parts of Asia hepatitis vaccines.

 Why do you need vaccines for typhoid when you travel to Africa?

 Where would you go to be vaccinated?

What are vaccines? 

A vaccine is usually a **weakened** form of the disease or a dead form of the disease. Vaccines may be given by injection or, as in the case of polio, in a drink. The person may be given an injection of antibodies, this will occur in the case of tetanus when the person is bitten by an animal or has soil enter a bad cut.

 Why might it be better to give the vaccine in a liquid rather than as an injection?

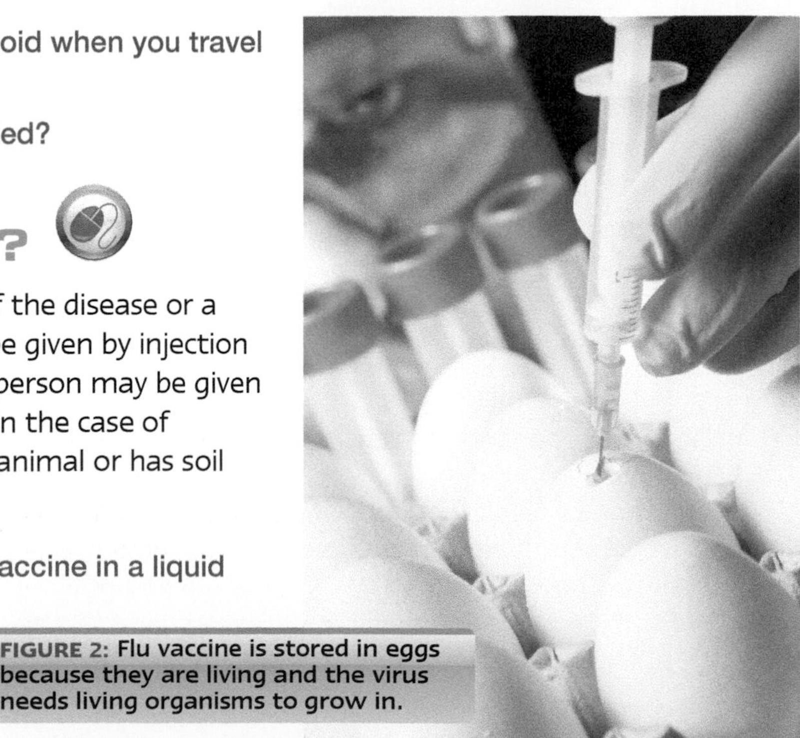

FIGURE 2: Flu vaccine is stored in eggs because they are living and the virus needs living organisms to grow in.

What do vaccines do?

- The vaccine enters the body and triggers the immune system to respond.
- White blood cells produce antibodies to attack the microbe.
- The microbe has been made harmless (it cannot reproduce) so it is destroyed quickly.
- The body now has defences against the microbe:
  Antibodies in the blood
  Special white blood cells called memory cells
- If the microbe then attacks the body, the memory cells produce antibodies to attack it quickly.
- In certain cases, people who are vulnerable (old or very young) are vaccinated if there is an outbreak of a disease such as the flu or meningitis.

How Science Works

Flu viruses keep changing. Every year scientists modify flu vaccines to match the latest flu strains. HSW

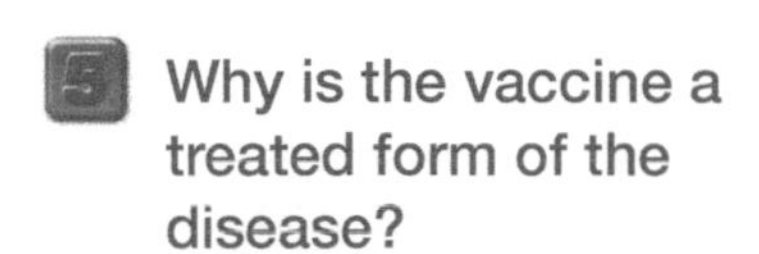

5 Why is the vaccine a treated form of the disease?

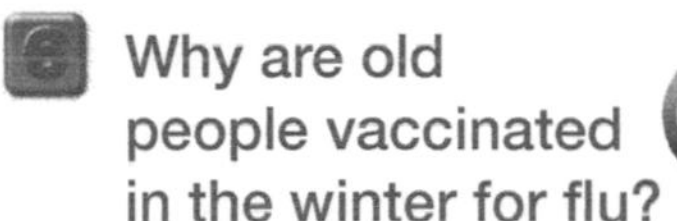

6 Why are old people vaccinated in the winter for flu?

Is there a risk?

Vaccines do not come without risk. **Side effects** can occur, and the doctor will inform you of the risks. In the case of whooping cough people who are susceptible to the disease can get brain damage, however, this is very rare. The vaccine for typhoid can make a person unwell for a while. In each of these cases the actual disease will have a worse effect.

FIGURE 3: Vaccines prevent the spread of disease in under developed countries.

Recently the vaccine for MMR (measles, mumps and rubella) was questioned by the papers and this led to a reduction of children receiving the vaccine. As a result there was an increase in the cases of measles with a few deaths from the disease. This happens because the more people who are vaccinated the less chance the disease can spread. This is called herd immunity.

7 Use the information above to give the case for and against the use of vaccines in children.

How to get rid of microbes

BIG IDEAS

You are learning to:
- Describe the difference between disinfectants and antiseptics
- Describe the use of disinfectants
- Design experiments to test the effectiveness of disinfectants 

Bleach cleans clean 

Adverts on the television often show 'magic' chemicals that kill all bacteria. These chemicals are usually **disinfectants** and bleaches. They have a special sign on the bottles to show they are toxic.

1. What does the sign in the picture mean?
2. a Where in the house do bleaches get used?
   b Why are bleaches kept out of young children's reach?

Disinfectants

Disinfectants work by killing the microbes. They can kill living cells so they should not come into contact with human tissue. **Antiseptics** can be used on humans because they stop the microbes growing. Some antiseptics are weakened forms of disinfectants.

3. a Why would you use a disinfectant not an antiseptic to clean the sink?
   b Why would you use an antiseptic on cuts?
4. Why have alcohol hand washes been introduced into hospitals?
5. Why are mops thrown away after they are used?
6. Why is most waste material from hospitals burned?

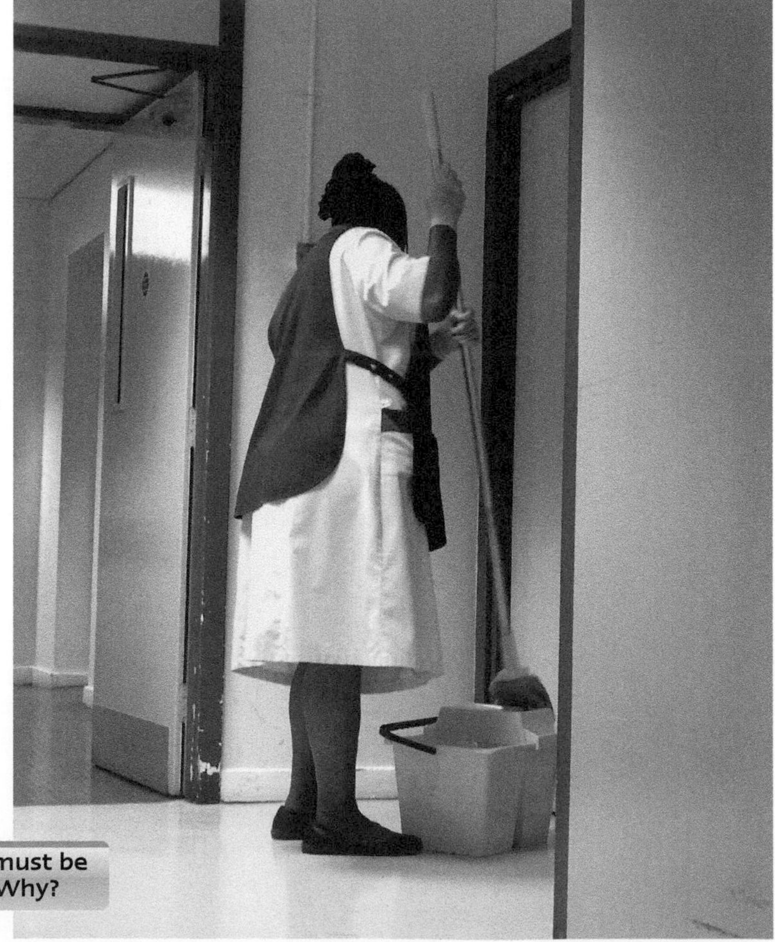

FIGURE 1: Hospitals must be constantly cleaned. Why?

Killing microbes

On bottles of disinfectants there is a recommended **dilution** for their use. If the dilution is too weak then microbes will not be killed and in very weak doses the microbes feed off the chemicals in the solution. Disinfectants work by killing microbes, they are used in hospitals to clean floors and sterilise equipment. Antiseptics are used on the skin. If you enter a hospital you will have to wipe your hands on alcohol rubs to remove bacteria.

FIGURE 2: An agar plate contains a jelly containing nutrients that bacteria thrive on.

Your task is to find out how effective disinfectants are.

You will be provided with an agar plate that is infected with bacteria.

You will be told how to dilute the disinfectant so you have a range of dilutions to use.

Method

Place a disc made from filter paper into each of the dilutions you have made. Take one out using tweezers and let the liquid drip off. Then place the disc on the agar plate at A and note the dilution. *Only lift the lid for a short time.*

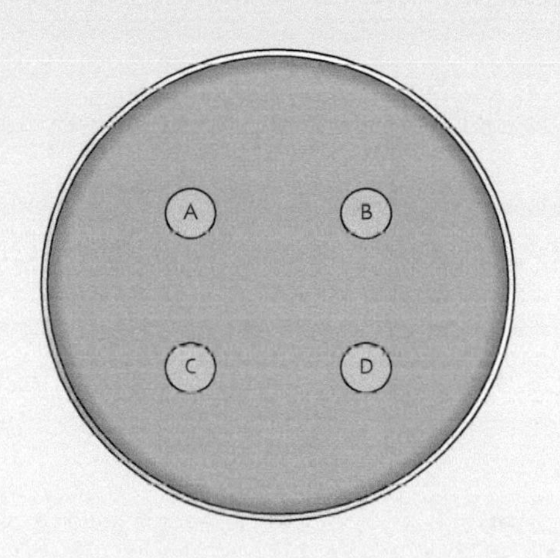

Repeat this for three more dilutions, placing each filter paper disc on the agar plate at B, C and D. Tape the plate shut and let it be incubated (placed in a warm container at 30 °C) for a few days. Now study the plate and measure the diameter of the clear area round each disc. Record your results in a table.

Questions

1. What range of dilutions have you used?
2. Why has the plate been taped shut?
3. Why is there a clear area round the discs?
4. Why is the plate incubated?
5. Explain your results, indicating what dose of disinfectants can be used and why.
6. What are the likely sources of **error** in the experiment?

How Science Works

Hospitals use agar plates to detect the presence of microbes in certain areas. The plate is left open in key areas and then incubated to see which microbes are present. Only fungi and bacteria can grow on the plates as viruses grow in living material.

7. Design an experiment to compare the effectiveness of these different disinfectants on a microbe.

Are microbes useful?

BIG IDEAS

You are learning to:

- Describe how microbes are used to make food
- Interpret information on the effect of temperature on yeast 
- Investigate the effect of temperature on the rising of dough 

What microbes have I eaten?

This morning I had **yoghurt** for breakfast. Then I ate some bread and **cheese** washed down with orange juice. At dinner time I had a burger. The burger was made from protein made by microbes. I have eaten many microbes, but I have not become ill.

FIGURE 1: What do microbes have to do with bread?

 Give the name of the foods made by microbes that I have eaten.

Making yoghurt

Yoghurt used to be made by leaving goat's or sheep's milk to curdle in the skin of an animal. Nowadays bacteria are added to the milk making the conditions acid (pH 4.5), this causes the milk to curdle. Flavouring is then added.

Cheese is also made using bacteria and a chemical called rennet. The milk separates into a more solid curd on the top and a liquid whey at the bottom. The whey is drained off and the curds are pressed to make the cheese.

 What is the difference between a soft cheese and a hard cheese?

 What types of food does cheese contain?

FIGURE 2: Why does Stilton have blue veins in it.

FIGURE 3: Why are some yoghurts described as 'live'?

Making dough

To make bread the fungus **yeast** is used. Flour is mixed with a solution of yeast. Sugar is added to provide a food source for the microbe to respire. Warmth helps the microbe multiply faster, so the **dough** is left in a warm area for several hours to rise. As there is a lack of oxygen the yeast respires anaerobically to produce carbon dioxide and alcohol. The carbon dioxide makes the bread rise. When the dough is put in the oven the alcohol evaporates off. Some breads such as pitta and naan bread do not rise because no yeast is added.

Breadmaking is an ancient art. In the Bible there is mention of unleavened bread, this is like pitta bread and is made without using yeast.

You are going to find out how temperature affects the rising of dough.

Method

1. Place five tablespoons of flour into your container.
2. Add 30 cm^3 of yeast solution to the flour and stir until it becomes a thick liquid that can be poured, add more flour or liquid as needed.
3. Carefully pour the liquid dough into three boiling tubes so they have equal amounts in. Measure the height of the dough once it has settled.
4. Place one boiling tube into a beaker of water at 20 °C, another in a beaker at 40 °C and the final one into one at 60 °C. Check the temperature in the beaker and change it to keep it near the right temperature.
5. After 30 minutes measure how much the height of the dough has risen.

How Science Works

Bacteria are grown on the waste material from certain food processing industries. The bacteria are used to make animal food.

Did You Know...?

Marmite is made from yeast and soya sauce is made using bacteria and fungi.

Questions

1. How have you made the test fair?
2. What pattern do your results show?
3. Explain your results.
4. Why is it difficult to draw a graph of the results?
5. Why do you not get drunk when you eat lots of bread?
6. Why are there holes in bread?

7. a) Explain how you could obtain more accurate results instead of measuring the height of dough.
   b) Explain why it is difficult to see a clear pattern with the results. What extra experiments would give a clearer pattern?

Practice questions

1 Copy the sentences below into your exercise book. Complete the following passage using the words below:

infection **symptoms** **droplets** **incubation period**

The measles virus enters the body by the spread of ___________ in the air. The virus enters the cells, this is called ___________. The virus may stay in the cells for a while until it starts to attack them, this is the ___________ ___________.

The person now starts to show the ___________, like a rash, sore throat and aches.

2 Smallpox is a disease that has been wiped out from Europe. It was once one of the biggest killers of people.

Year	Deaths of vaccinated people %	Deaths of non-vaccinated people %
1901	10	36
1902	10	34
1903	3	5
1904	5	8

a What pattern is shown by the table?

b What evidence is there that vaccinations prevented smallpox?

c What evidence is there that as more people are vaccinated the number of deaths from smallpox of all people is reduced?

d Why should this be?

e Explain how the vaccine causes the body to produce antibodies.

3 A student tried to find out which disinfectant was the most effective against a certain bacteria.

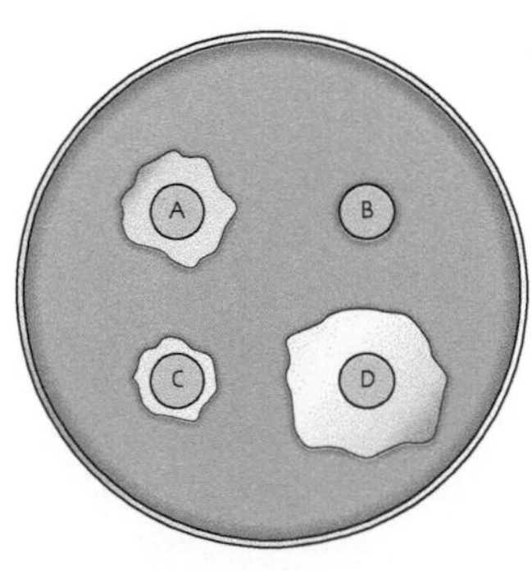

He made some agar plates and placed the bacteria on them. He then placed discs of four different disinfectants on the plate. He incubated the plate at 30 °C and looked at the results.

a Why is there a clear zone round the discs?

b How can you measure the zone round the discs?

c Which disinfectant is the most effective? Explain why.

d Why is there no zone round B?

e Give **one** way you will make the test fair for comparing the effect of the disinfectants.

4 A dangerous disease breaks out on a ship. The ship is put into quarantine.

a Why is the ship put into quarantine?

b What precautions will scientists take when they board the ship?

c The scientists found a sample of a microbe and wanted to find out if it caused the disease.

i How could they use mice to see if the microbe caused the disease?

ii Why might the use of mice be unacceptable?

5 Measles, mumps and rubella are given as a triple injection. Several years ago there was concern expressed about the possibility that the vaccine could cause autism.

In one study, 75% of a very small sample suffered from autism after they had the MMR jab. In another study of 500 children with autism, half of them had developed autism before vaccination.

a What evidence is there that MMR could cause autism?

b What evidence is there that MMR is unlikely to cause autism?

c Why do scientists publish their findings through scientific papers before they make the public aware of them?

The publicity in some of the papers resulted in parents getting very worried and not giving their child the MMR vaccine. This resulted in a measles epidemic in America with 120 children dying.

d Why did the measles epidemic occur?

6 Pasteur was famous for his work on microbes.

He trialled a vaccine for anthrax on sheep in the following way:

- He took 60 sheep for the trial.
- 10 sheep were left alone in the field.
- He then split the remaining 50 and injected 25 with a weak form of anthrax.
- He then injected these 25 with a strong form of anthrax and the ones he had not treated with a strong form of anthrax.

a What results would you expect if the vaccine worked?

b Why were 10 sheep not treated?

c Why did he use 25 sheep not five sheep?

d He vaccinated the 50 sheep with the strong vaccine alternately (i.e. one of the untreated ones followed by one of the treated ones). Explain why.

Learning Checklist

4

- ☆ I can name three ways diseases are spread. page 34
- ☆ I know that infectious diseases are caused by microbes. page 35
- ☆ I know two ways the body fights off infection. page 36
- ☆ I know one example of biological warfare. page 42

5

- ☆ I know that the types of microbe that cause disease are bacteria, viruses and fungi. page 34
- ☆ I know two examples of viral and two of bacterial disease. page 38
- ☆ I can explain how one disease is spread and the symptoms it has. page 40
- ☆ I know that disinfectants kill microbes. page 48
- ☆ I can explain how to use an agar plate to show how effective disinfectants are. page 49

6

- ☆ I know the infection cycle explains how microbes cause disease. page 34
- ☆ I can explain the difference between infection and symptoms. page 35
- ☆ I know how white blood cells prevent disease. page 36
- ☆ I know the role of antibodies. page 37
- ☆ I know the difference between disinfectants and antiseptics. page 48

7

- ☆ I can explain how vaccines work. page 47
- ☆ I can explain the case for and against the use of vaccines. page 47
- ☆ I can explain how to find out which strength of disinfectant is not effective against bacteria. page 49

8

- ☆ I can explain the evidence to support the germ theory for disease. page 33
- ☆ I can use the immune response model to explain how allergies are caused, how vaccines work and why some people never catch a disease. page 44

Topic Quiz

1 What did people think caused diseases in 1600?

2 Who discovered the process of pasteurisation?

3 Name a food produced using microbes.

4 How is a cold spread?

5 How is malaria spread?

6 Name **two** ways the body stops diseases entering.

7 Name **two** types of microbe.

8 What is the role of white blood cells in preventing disease?

9 What is meant by the incubation period?

10 What is the difference between symptoms and infection?

11 What does a microbe do to the body to cause disease?

12 How does a vaccine prevent a person getting a disease?

True or False?

If a statement is false then rewrite it so it is correct.

1 All microbes cause disease.

2 Dead bodies were catapulted into castles in ancient times to spread disease.

3 White blood cells produce antibodies.

4 Infectious diseases can be spread.

5 HIV is the most common STD.

6 The MMR vaccine is for measles, mumps and rubella.

7 HIV is the virus and AIDS is the symptoms caused by the virus.

8 Bread rises because yeast expands.

9 There are over twenty different sexually transmitted diseases.

10 Flu is a sexually transmitted disease.

Literacy Activity

Read the passage below then answer the questions.

Louis Pasteur was one of the great scientists. He was responsible for the 'Germ Theory'; this stated that microbes caused diseases.

A colleague's wine was going sour (turning to vinegar) at a local vineyard. Pasteur investigated and discovered that microorganisms had contaminated the wine. He developed a process to kill the bacteria and not ruin the taste, this was called pasteurisation.

He then investigated why silkworms were dying in silkworm nurseries in France.

To do this he:

- studied the silk moths' cells and found microorganisms.
- injected fluid from the moths into healthy moths.
- found the silk moths died and had the same microorganisms in their cells.

1 What does sour wine form?

2 What food material is pasteurised?

3 What is the benefit of pasteurising milk over boiling it?

4 How did Pasteur's work on silk worms provide evidence to support the Germ Theory?

FIGURE 1: Salt city underground.

BIG IDEAS

By the end of this unit you will be able to use the particle model to explain what happens when solids dissolve and how mixtures can be separated. You'll be able to plan and carry out experiments to test your ideas.

Taking salt from underground

When seas dry up, salt is left behind. There are layers of salt hidden underground all over the world. Salt has always been valuable to flavour and to preserve food. We can make useful chemicals from salt too. The chemical name for salt is sodium chloride. We can make chlorine gas from salt. Chlorine is used to make drinking water safe by killing harmful bacteria and also for swimming pools.

In winter you often see lorries spreading salt and grit on the roads. The salt makes the ice melt and the roads are safer for drivers. This kind of salt is called rock salt. It is a natural mixture of salt, clay and dirt. It is fine for the roads but not pure enough for cooking or for using in a dishwasher. To obtain pure salt we need another kind of salt mine.

In Cheshire people have obtained salt from brine pits since the seventeenth century. Brine is just salty water. They realised that there must be layers of salt buried underground. Since salt is soluble in water but rocks are not, we can use water to mine the salt for us. We call this solution mining.

It is very simple. Water is pumped down one of the pipes and it dissolves some salt. The brine is then pumped up again and evaporated to leave pure salt. There is just one little problem. Removing the salt from under the ground leaves large holes. The land above can sink or collapse into these holes, destroying buildings and walls. You can see the effects of this subsidence all over the Cheshire salt fields.

FIGURE 2: Salt can melt the ice.

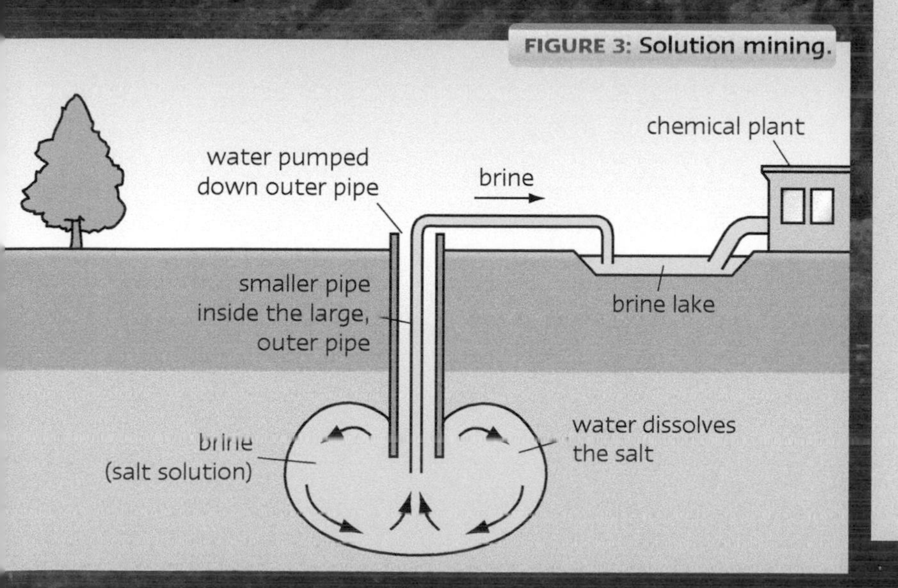

FIGURE 3: Solution mining.

What do you know?

1. Give **two** uses for salt.
2. How does swimming in the sea prove that salt dissolves?
3. Where in Britain can you find underground salt?
4. Why do lorries spread salt on the roads?
5. What is rock salt?
6. What do we mean by solution mining?
7. Name **one** chemical that is made from salt and explain how it is used.
8. Why did people describe a visit to the seaside as 'going down to the briny'?
9. What features in the landscape might show that solution mining has taken place?
10. Why do people say that it is impossible to drown in the Dead Sea?

Dissolving rocks

BIG IDEAS

You are learning to:
- Recognise that rocks can dissolve
- Provide evidence that water is a good solvent
- Explain the role of calcium carbonate in petrifaction

5

Disappearing rocks 

In some places streams simply disappear underground. The water has already dissolved some of the limestone rock to make a hole in the ground called a swallow hole. Limestone caves are amazing places. There are **stalactites** growing down from the cave roof. You can often see water dripping from the ends of the stalactites. Where these drips hit the cave floor, **stalagmites** grow upwards. They grow rather slowly, a few centimetres in a hundred years.

We know that the water has dissolved the rock by looking at the evidence. If you collect the cave water and boil it, it leaves **limescale** behind in the kettle. This is the same chemical as in the rock that was dissolved by rainwater. We also find limescale building up around taps in the limestone areas of the country.

1. What is a swallow hole?
2. How quickly do you think limestone dissolves? 
3. Give **one** piece of evidence that water can dissolve rocks. 

FIGURE 1: Even rocks dissolve.

Petrifying

If you have a favourite toy from childhood or pair of shoes, you can preserve them for ever. Petrifying springs have water that contains **calcium carbonate**, the chemical that forms limestone. If you leave a soft toy in the spring for six months, it turns to stone. It becomes **petrified**. As the spring water evaporates, it leaves behind crystals of calcium carbonate. This is how stalactites grow as well.

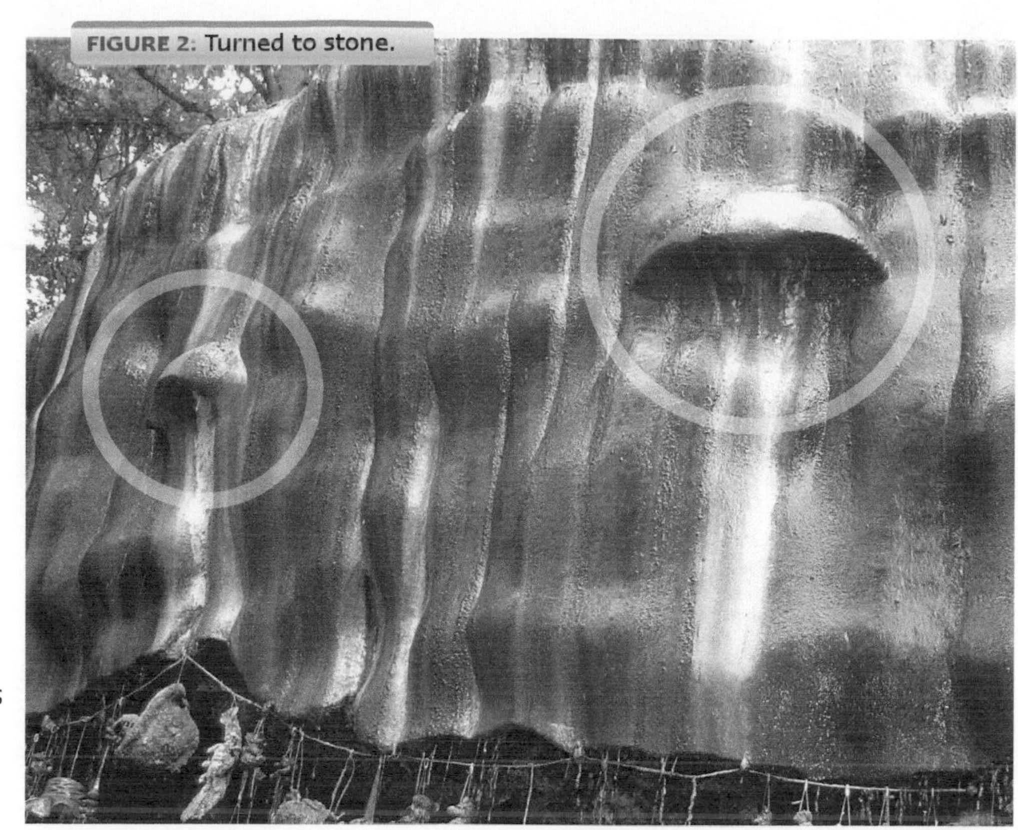

FIGURE 2: Turned to stone.

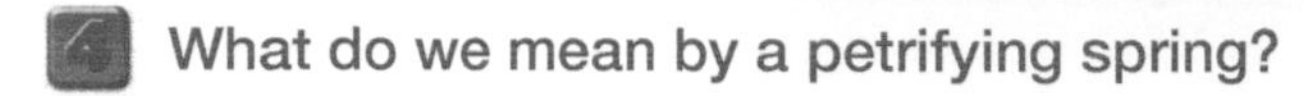

4 What do we mean by a petrifying spring?

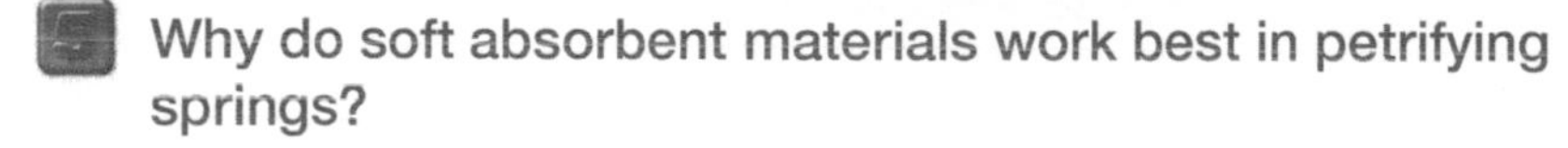

5 Why do soft absorbent materials work best in petrifying springs?

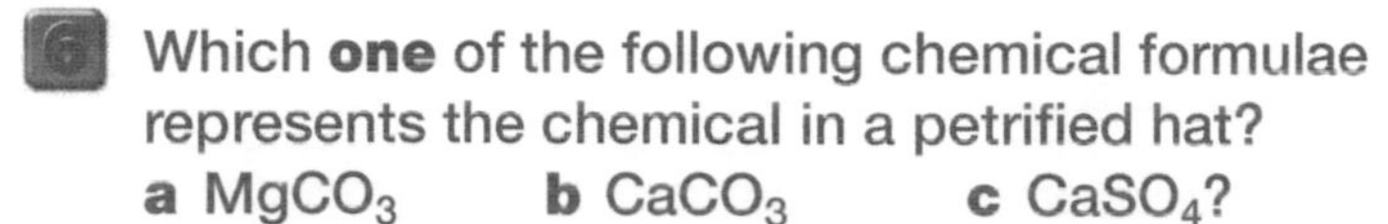

6 Which **one** of the following chemical formulae represents the chemical in a petrified hat?
a $MgCO_3$ **b** $CaCO_3$ **c** $CaSO_4$?

FIGURE 3: Natural spring water.

Did You Know...?

If you are petrified, it means that you are too frightened to move. It means that you look as though you have turned to stone. This is the same word we use for petrifying springs and wells. People who study the way rocks form are studying petrology.

7 Design an experiment to find out whether wool, cotton or nylon fabric is the most absorbent (for example, of spring water). Make clear any measurements you need to make. HSW

Sweet tooth

BIG IDEAS

You are learning to:
- Describe the way sugar dissolves
- Discuss how the solubility changes
- Evaluate the evidence for the effect of temperature on solubility

Sugar in my tea

Some people can't drink their tea or coffee without sugar. They add spoonfuls of sugar and stir. The sugar crystals disappear, they dissolve in the water. We know how much sugar has dissolved by the taste. The sweeter the taste, the more sugar has dissolved.

1. Why does sugar disappear when you stir it in tea?
2. How do you know that the sugar is still there in the drink?

Shape up

If you use a glass beaker instead of a mug, you can see just how easily the sugar dissolves. There are many different kinds of sugar that we can buy. Some are fine powders such as caster sugar, others are big lumps of sugar. They all contain the same chemical compound, called **sucrose**. It is extracted from sugar cane or from sugar beet in Europe. The sucrose particles (molecules) are the same in each type of sugar.

FIGURE 1: All the sugars.

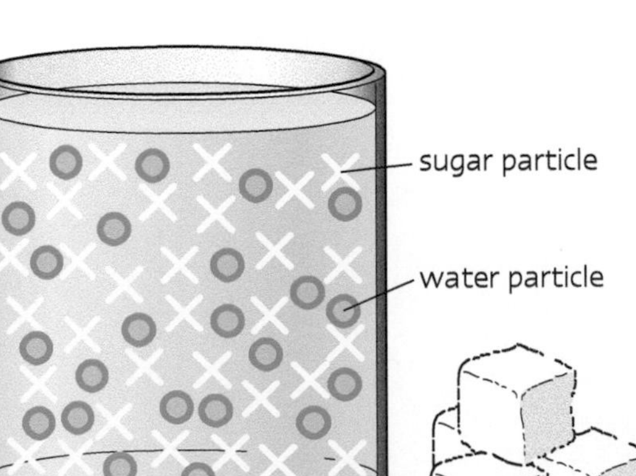

FIGURE 2: Sugar dissolving in water.

Not all types of sugar dissolve in water at the same speed. Even if you make it a fair test by having equal masses of sugar in equal volumes of water, the results vary. Look at these results.

Sugar type	Mass used (g)	Volume water (cm^3)	How many stirs to dissolve it all
Caster	1.50	100	20
Lump	1.50	100	75
Granulated	1.45	100	42
Coffee lumps	1.50	100	58

3 Which kind of sugar was the hardest to dissolve?

4 Why did the sugar that dissolved quickest disappear so fast?

5 Put the kinds of sugar in order of size of the sugar pieces, starting with the largest.

Temperature effects

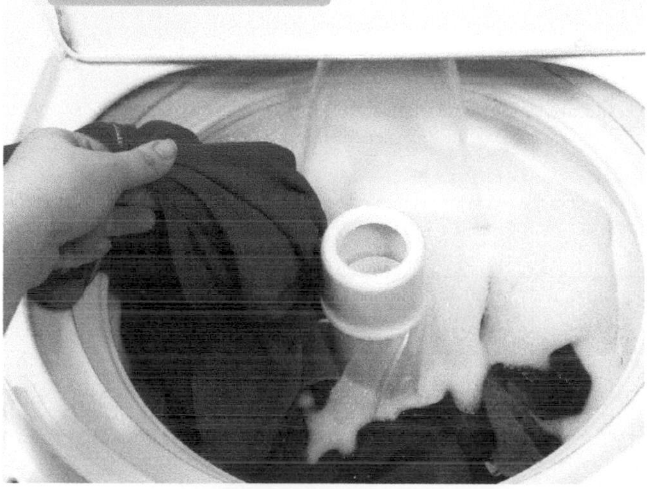

One way to help things dissolve is to use hot water. This is why we wash clothes in warm water not cold water. The clothes don't dissolve but some of the dirty stains do. Of course, we also add detergent to help with the cleaning process. Any **soluble** materials in the clothes will dissolve more at a higher temperature. Water particles (molecules) must come into contact with a solid, such as sucrose, to allow it to dissolve. The sucrose particles can then spread out in the solvent. We say that their solubility will change with temperature. Look at the data from an experiment to dissolve sugar in water at different temperatures. In each case, the results show the mass of sucrose that can dissolve in 100 cm^3 of water.

FIGURE 3: Heat helps.

Temperature of water (°C)	0	20	40	80
Mass of sucrose that can dissolve (g)	180	200	240	600

6 Estimate the mass of sucrose that will dissolve at 60 °C.

7 How could you display this data better?

8 What is the link between solubility and temperature, as shown by the data for sucrose?

9 Explain the connection between the energy of the water particles at higher temperatures and the resulting solubility of sucrose.

Did You Know...?

Metals can dissolve in each other. The result is an **alloy** such as bronze or brass. Dentists used to use mercury alloys for fillings in teeth. One was called Sullivan's Amalgam, made from copper dissolved in mercury. Now most fillings are made from ceramics that avoid the possible harmful effects of metal alloys in your mouth.

Pure salt

BIG IDEAS

You are learning to:
- Explain how salt is obtained
- Use an understanding of solubility to make pure salt

Mining salt

The salt from mines is very dirty, it is not **pure** salt. This rock salt varies a lot in colour. Most rock salt is brown but it can be yellow or even red. The colour comes from clay mixed with the salt. Salt is **soluble**, it dissolves in water. Clay is not soluble. The rock salt from the mine is broken into a powder. This makes it easier to spread the salt on icy roads in winter.

1. What do we call the kind of salt from a salt mine?
2. Why isn't this mined salt pure?

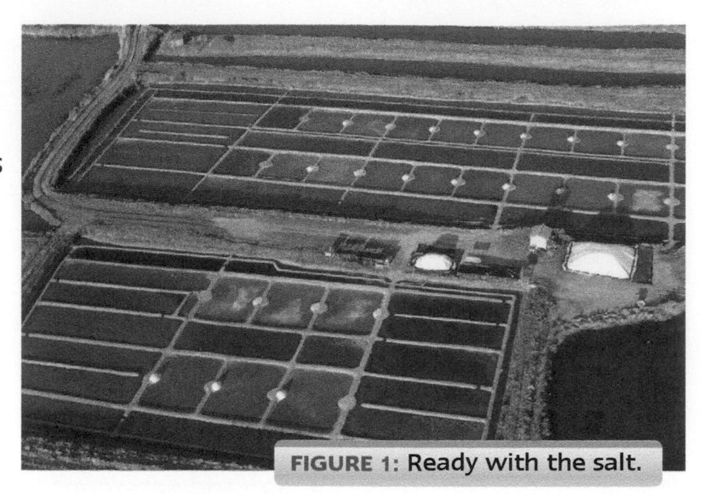

FIGURE 1: Ready with the salt.

Purifying salt

People have obtained salt from seawater for thousands of years. The Romans in Britain trapped seawater in shallow ponds. As the water evaporated, salt crystals were left behind. The same method is still used today in countries such as France and Australia. The seawater is taken from an area that is not polluted. The sea salt crystals are pure enough for us to use in cooking, unlike rock salt. Rock salt contains **insoluble** particles, including sand and clay. These must be removed before the salt can be used with food.

3. How did the Romans obtain salt in Britain?
4. What do we mean by an insoluble material?
5. What causes the colour in rock salt?

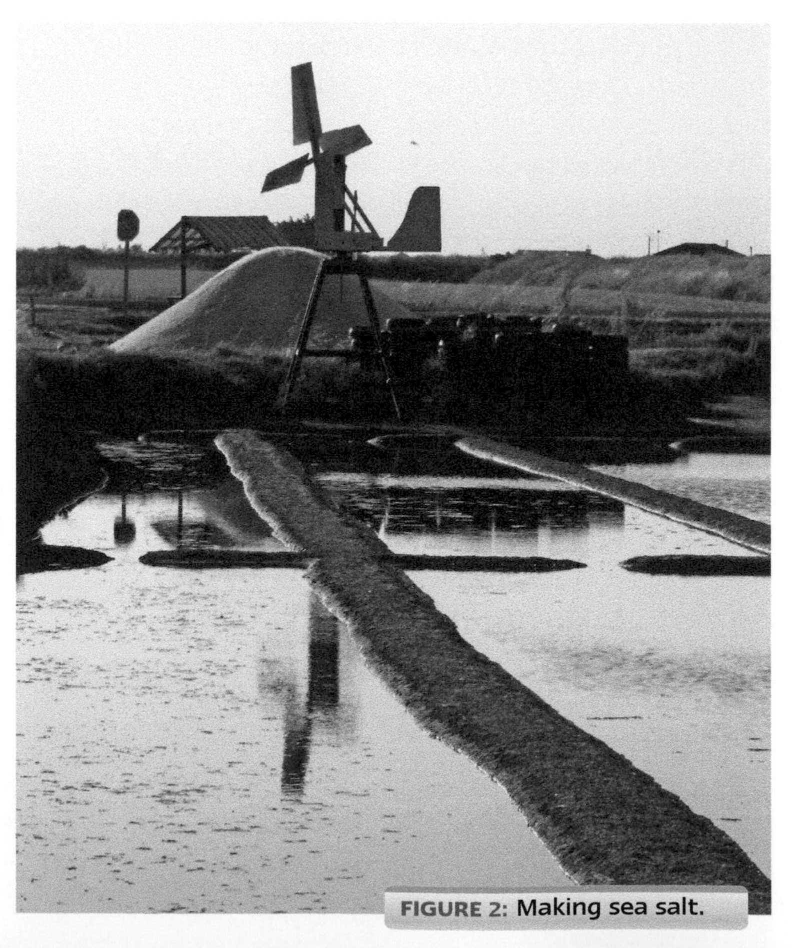

FIGURE 2: Making sea salt.

... filtered ... insoluble ... pure

Making pure salt

You can purify rock salt yourself. Since the salt is soluble but the sand is not, we can use water to help. Insoluble materials can be **filtered** off using a funnel and filter paper.

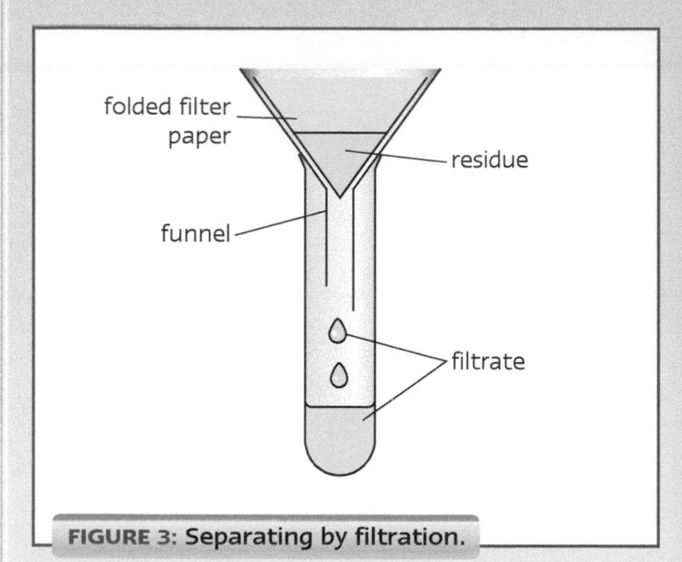

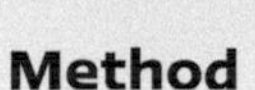
FIGURE 3: Separating by filtration.

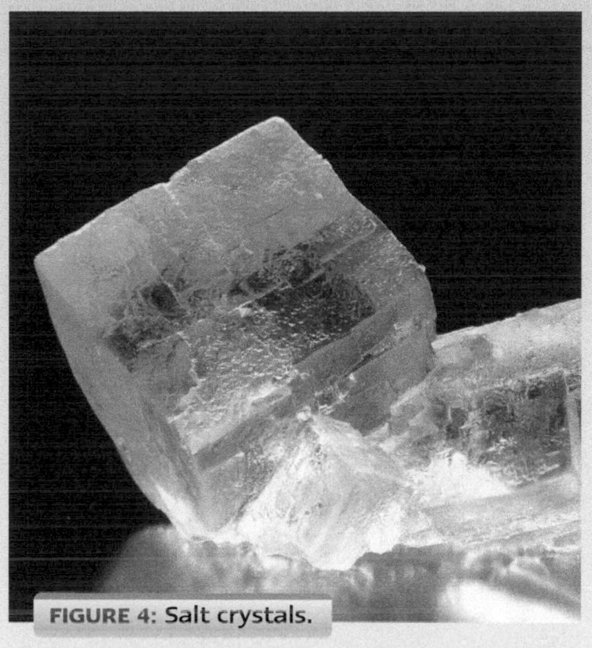
FIGURE 4: Salt crystals.

Method

1. Measure 100 cm^3 water into a beaker.
2. Add 6 spatulas of crushed rock salt.
3. Stir for at least 5 minutes to dissolve the salt.
4. Filter the mixture.
5. Collect the solution in a clean beaker.
6. Half fill an evaporating basin with this solution.
7. Heat gently over a Bunsen until most of the water has evaporated.
8. Stop before it becomes dry or it will start 'spitting' salt crystals at you.
9. Leave to cool and examine the salt crystals.

How Science Works

A **solvent** is a liquid that dissolves other materials. Water is a good solvent for purifying salt. If salt were insoluble in water, scientists would need to try a range of different solvents instead. Other solvents include oil, alcohol and nail polish remover (acetone, also called propanone). HSW

Questions

1. What was the insoluble material?
2. Describe and compare the appearance of the rock salt and the purified salt.
3. Why might the purified salt still not be safe to eat?

4. The rock salt contains both particles of salt and particles of sand and clay. Use the particle theory to explain your observations on stirring the salt with water.

Super solvents

BIG IDEAS

You are learning to:
- Sort out different solvents
- Compare different solvents
- Recognise the nature of hazards associated with solvents 

Modern art

The pictures and letters that you see on buildings and trains are called **graffiti**. They are quick to do but very hard to remove. Most graffiti artists use spray paints. Water cannot dissolve spray paint. You cannot wash away the pictures with water.

1. What do we mean by graffiti?

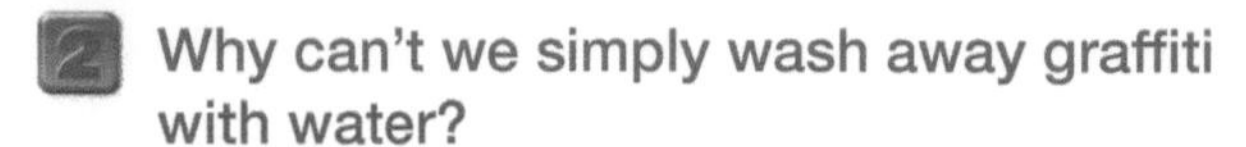

2. Why can't we simply wash away graffiti with water?

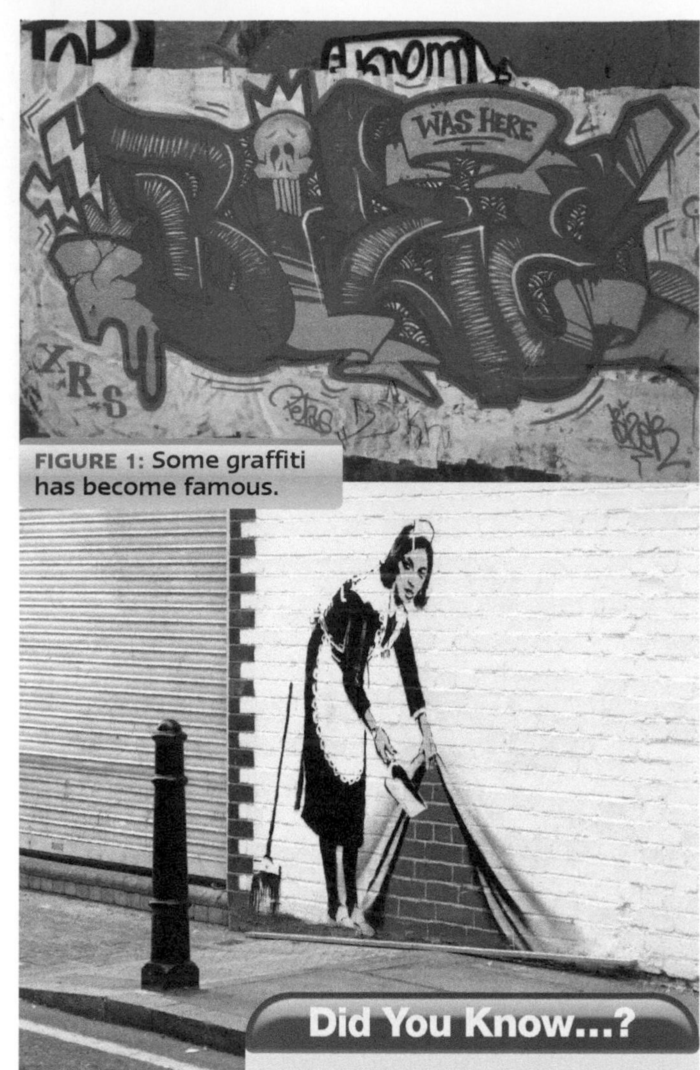

FIGURE 1: Some graffiti has become famous.

Solvent choice 

Water has been called the universal solvent, meaning that it can dissolve anything. Luckily this is not true. Clearly most of the solutions we use in a laboratory are solutions in water. Almost all acids and alkalis are soluble in water, as are salts such as common salt (sodium chloride). Many materials are **insoluble** in water. There are other solvents available that we can try instead. For example, petrol is a very good solvent for stains but is too dangerous to use since it is so flammable.

Solvent	It can dissolve	Does the solvent burn?
Water	Sugar, food colours, emulsion paint	No
Alcohol	Ballpoint pen ink, perfume, herbs and spices	Yes
Acetone (propanone)	Nail polish	Yes
White spirit	Grease, oil paint	Yes

Did You Know...?

Many perfumes and liqueur drinks contain alcohol. The alcohol can dissolve the colours, flavours and odours to make the product. Alcohol evaporates easily. This is why perfume or aftershave dries so quickly on your skin.

FIGURE 2: Making perfume.

3 Which solvent would you choose to remove these things?

a Nail polish from glass.

b Ballpoint pen mark from a shirt.

4 What hazard might there be in using solvents other than water?

How Science Works

Comparing solvents

You are going to see how a range of solvents can be compared.

Method:

Your teacher will give you a piece of white-painted board with marks on it from a pencil, marker pen, fountain pen and ballpoint pen.

You are going to moisten a cotton wool bud with one of the solvents provided and then draw it across the marks on the boards. The solvents you are given could include water, alcohol and nail-polish remover.

Note any changes made to the marks. Do any of them disappear?

Record your observations.

Discuss with a partner which one is the best solvent and why.

HSW

Tar, tar

Road surfaces eventually wear out and must be replaced. On minor roads, this is often done by spreading hot tar followed by rock chippings. If you drive over the new surface you get tar stuck to the paint of your car. Tar is insoluble in water and difficult to remove. If you try the solvent acetone, there will be problems. The acetone will remove the tar but may take off the paint too. The choice of solvent is very important and must be done by careful testing and checking.

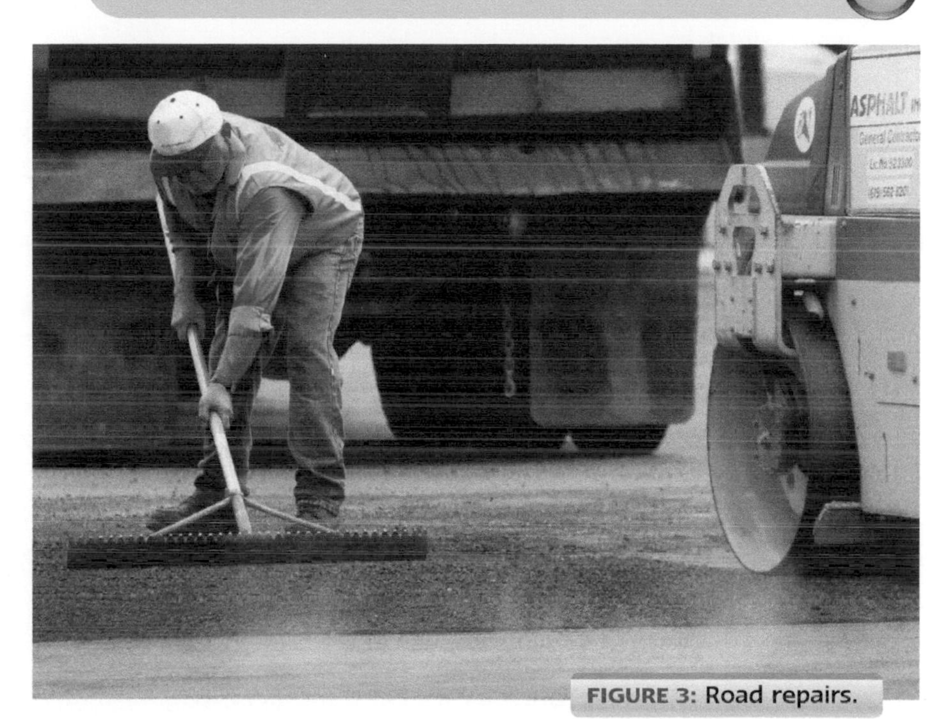

FIGURE 3: Road repairs.

5 Why might there be a problem in removing tar from a car?

6 What other hazard is associated with driving over a newly surfaced road where rock chippings were used?

7 Emulsion paint is used for ceilings and walls. How would you clean the paint brushes afterwards?

8 Design an experiment to compare how **three** different white paints are affected by the rain and weather. Decide on the best location for the paint samples and how long such a test should last.

How Science Works

Scientists test solvents at a range of temperatures. Some materials dissolve well in a hot solvent but not at lower temperatures. Care must be taken to condense solvent vapour to avoid the danger of a fire. Good ventilation is also important.

HSW

Distillation

BIG IDEAS

You are learning to:
- Explain what happens when liquids are separated
- Discuss distillation in terms of changes of state 

Heating up 

On a cold day, the steam from a hot bath or from a boiling kettle, can **condense** on a cold surface. The water vapour cools down on a window or wall and turns back to water again. The changes are like this:

Hot water ⟶ steam ⟶ water again

This is also what happens in **distillation**. We can separate liquid mixtures using distillation.

FIGURE 1: Escaping steam.

1. Why does steam turn into water on a window?
2. What do we mean by the word condense? How could you show it by using a mirror?

Distilling mixtures

FIGURE 2: Distilled beauty.

How Science Works

The distilled water can be tested to see if it still contains salt. The chemical called silver nitrate turns white if there is salt present. Distilled water should be pure water and give a **negative result**, no white colour. Scientists develop new tests all the time to find out if a particular material is present or not.

Distillation is widely used in making perfumes, fuels such as petrol, and alcoholic drinks such as vodka. There are two changes of state involved in distillation. First some liquid must be evaporated by heating. Then the vapour is cooled and condenses back to a liquid. Not all materials can be distilled, for example, salt in seawater. When salty water is heated, only the solvent, (the water) changes state. The solid salt is left behind since its boiling point is much higher than that of water. Sand behaves in the same way as salt. Distilling sandy water gives pure water. We call it distilled water and use it to fill steam irons. Different liquids boil at different temperatures. This helps us to separate mixtures of liquids.

3. What are the changes of state in distillation?
4. What is distilled water and how do we use it?
5. Name **one** product we buy that has been distilled.

Purifying seawater

You can separate pure water from seawater using distillation. The same change happens in nature when the Sun evaporates water from the ocean. The clouds and the rain are fresh water, not salt water.

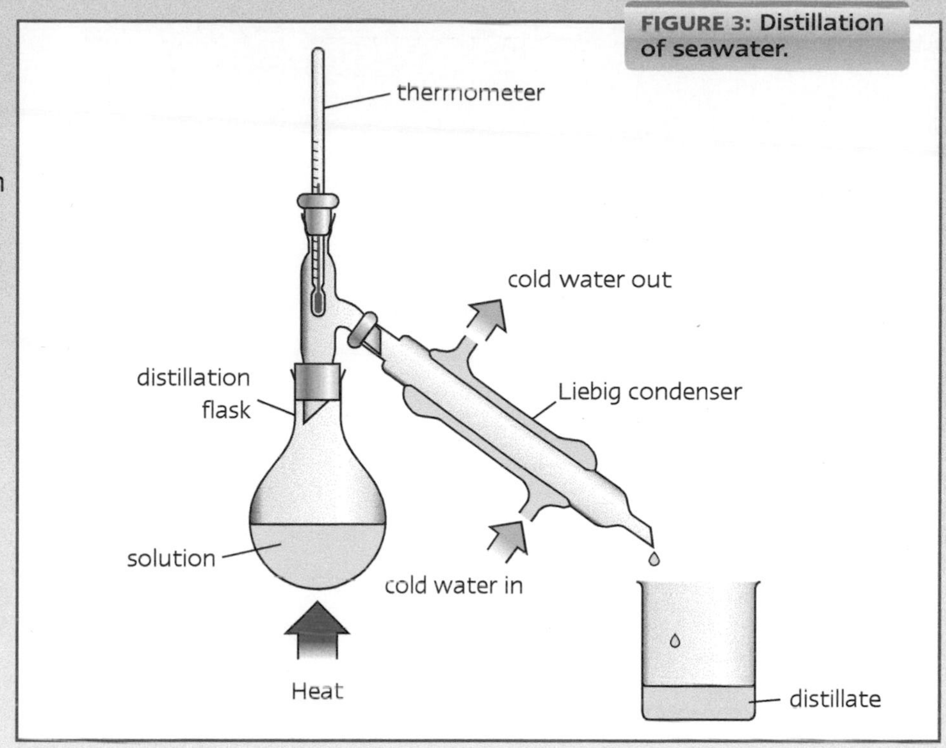

FIGURE 3: Distillation of seawater.

Method

1 Set up the apparatus as shown in the diagram.
2 The flask should be less than half full.
3 Place an empty tube or beaker to collect the condensed steam. Support this tube in a test-tube rack.
4 Heat the flask of seawater using a Bunsen.
5 Collect half a tube of condensed steam.
6 Pour the distilled water into an evaporating basin and leave to evaporate. Do the same to a sample of the original seawater for comparison.
7 When the water has all evaporated, compare the two.

Questions

1 Does all of the steam condense in the tube?
2 How do you know if the distilled water still contains salt?
3 How could you improve this experiment and avoid losing steam into the air?

4 There are forces of attraction between the particles in liquids. Explain why raising the temperature causes a liquid to boil.

Better boiling

BIG IDEAS

You are learning to:

- Decide how to improve distillation
- Recognise that distillation has a long history
- Explain how a special condenser improves performance

Catching steam

When you boil water and then cool the steam again, it is just exactly what happens in distillation. It is hard to catch all of the steam. Steam is a gas and it just disappears into the air. To catch the steam and stop it escaping, the trick is to cool the steam down. Then the steam turns back into water again.

1 Why does the steam turn back to water on the frozen peas?

2 Does using frozen peas catch all of the steam?

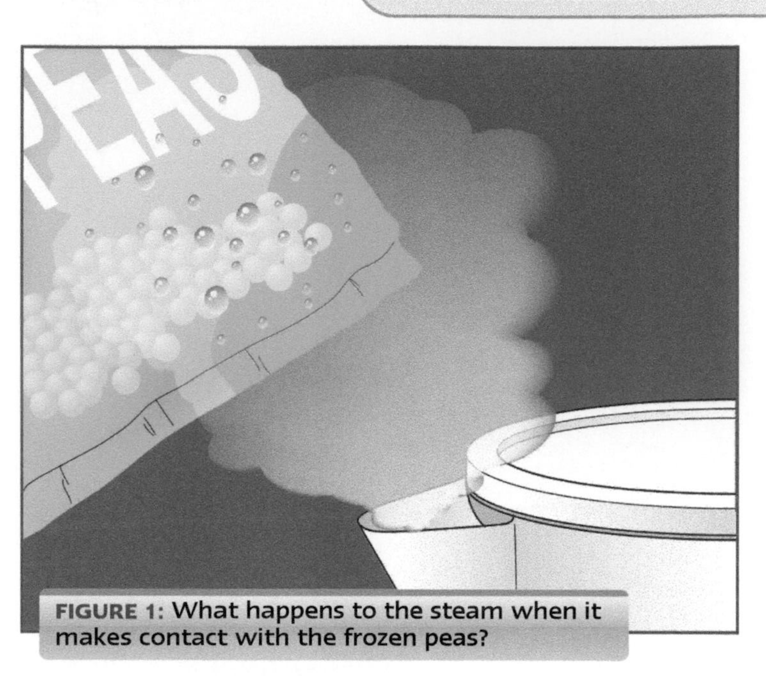

FIGURE 1: What happens to the steam when it makes contact with the frozen peas?

The alchemists

The early scientists in Greece two thousand years ago were known as **alchemists**. They invented a way to distil liquids. They designed a piece of apparatus which they called a still, because it was designed to distil liquid mixtures. It was so successful that the design was still advertised for sale in science catalogues of 1860. The problem with distillation is how to turn all of the hot **vapour** back into a liquid. A simple distillation experiment using just a glass tube as a condenser, rather than a Liebig one, loses most of the steam. The alchemists' still trapped the hot vapour, giving it a chance to cool and start to condense. It was made of copper. Copper is a metal that is very **malleable**, which means it is easy to bend into complicated shapes. Copper does not **corrode** easily, unlike iron which goes rusty. Copper also conducts heat well. It was an ideal material for the still.

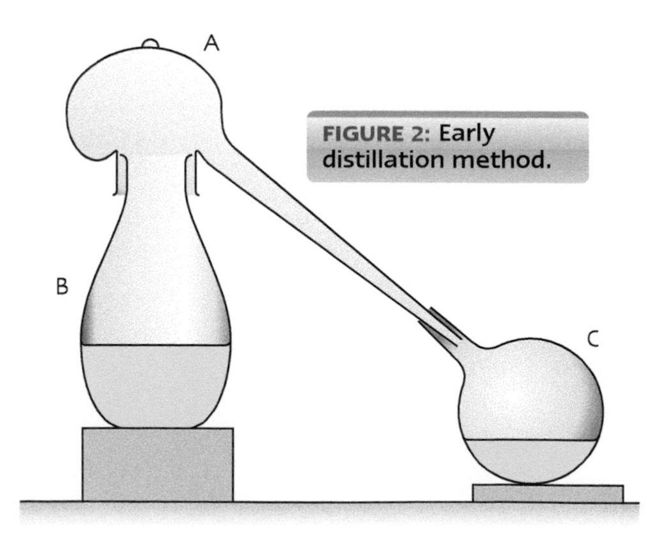

FIGURE 2: Early distillation method.

3 What was the name of the early scientists?

4 Give **two** reasons why copper is a good choice for a distillation apparatus.

5 What is the main problem with using a simple distillation apparatus, a glass tube?

... alchemist ... corrode

Distillation today 

The glass distillation apparatus that we use today is based on the same principle as that of the alchemists. Look back at page 67 for a labelled diagram of this apparatus. The major improvement is in the design of the condenser. The Liebig condenser is a double glass tube, one inside the other. The hot vapour from the boiling liquid flows through the inner tube. The separate outer tube contains a stream of cold water to cool and condense the hot vapour. The cold water enters the condenser at the lower end. The condenser must be full of cold water before any can overflow to the sink. This improved design will condense most vapours easily. The liquid that is collected at the end is called the **distillate**.

How Science Works

The history of the condenser shows how science improves. The problem of cooling and collecting the distillate was solved by changing the design. Not all design changes are so successful. The failures are useful too. We learn from our mistakes in science. HSW

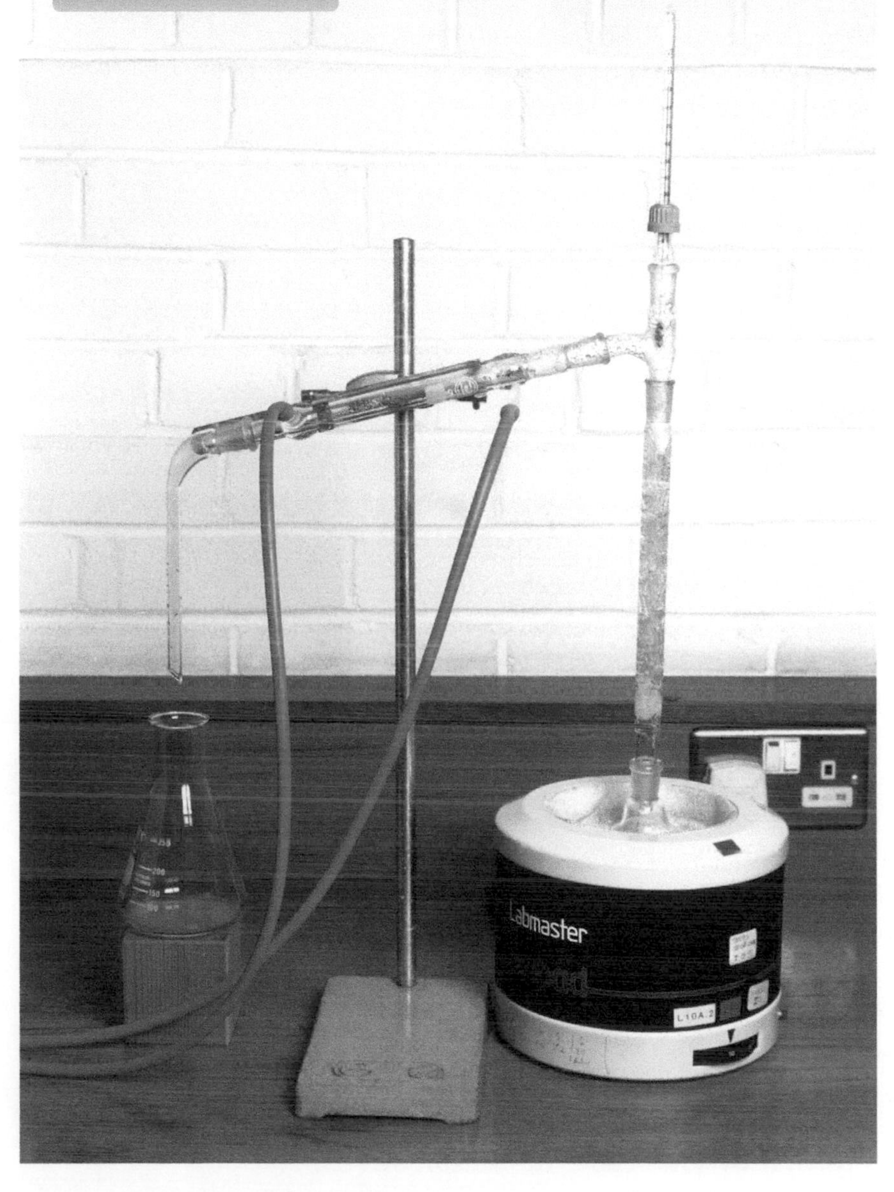

FIGURE 3: Laboratory distillation.

How Science Works

6. Why does the Liebig condenser need a double tube?
7. Why is this condenser more efficient than a simple glass tube? HSW
8. What modern scientific words derive from the same word as alchemist?

9. Give **two** reasons to explain why perfume manufacturers choose to use an electrically heated distillation apparatus.

Major crime solved

Police scientists have today smashed a major crime gang. The gang had terrorised a large area of the city for over a year. A series of burglaries and break-ins at jewellery stores had netted the criminals thousands of pounds. Unfortunately for them they were careless, leaving behind clues to their identities wherever they went. The gang always relied on a sports car to make their escape. Fast driving causes accidents and the driver left traces of car paint at several of the crime scenes. One of the robbers was so confident that he left messages scrawled on the wall, taunting the police to catch them. Well today, his wish came true. The robbers thought they were too clever and too fast to be caught. But they reckoned without the super sleuths in the police forensic team. Using chromatography, the team proved in court that they had arrested the right gang. The paint left at the scene matched their own car perfectly, even though they had repaired the scratched paintwork. The leader of the gang who left the messages made a big mistake. Chromatograms of the ink from the wall exactly matched that found in his own set of marker pens. Another good day for detectives and science.

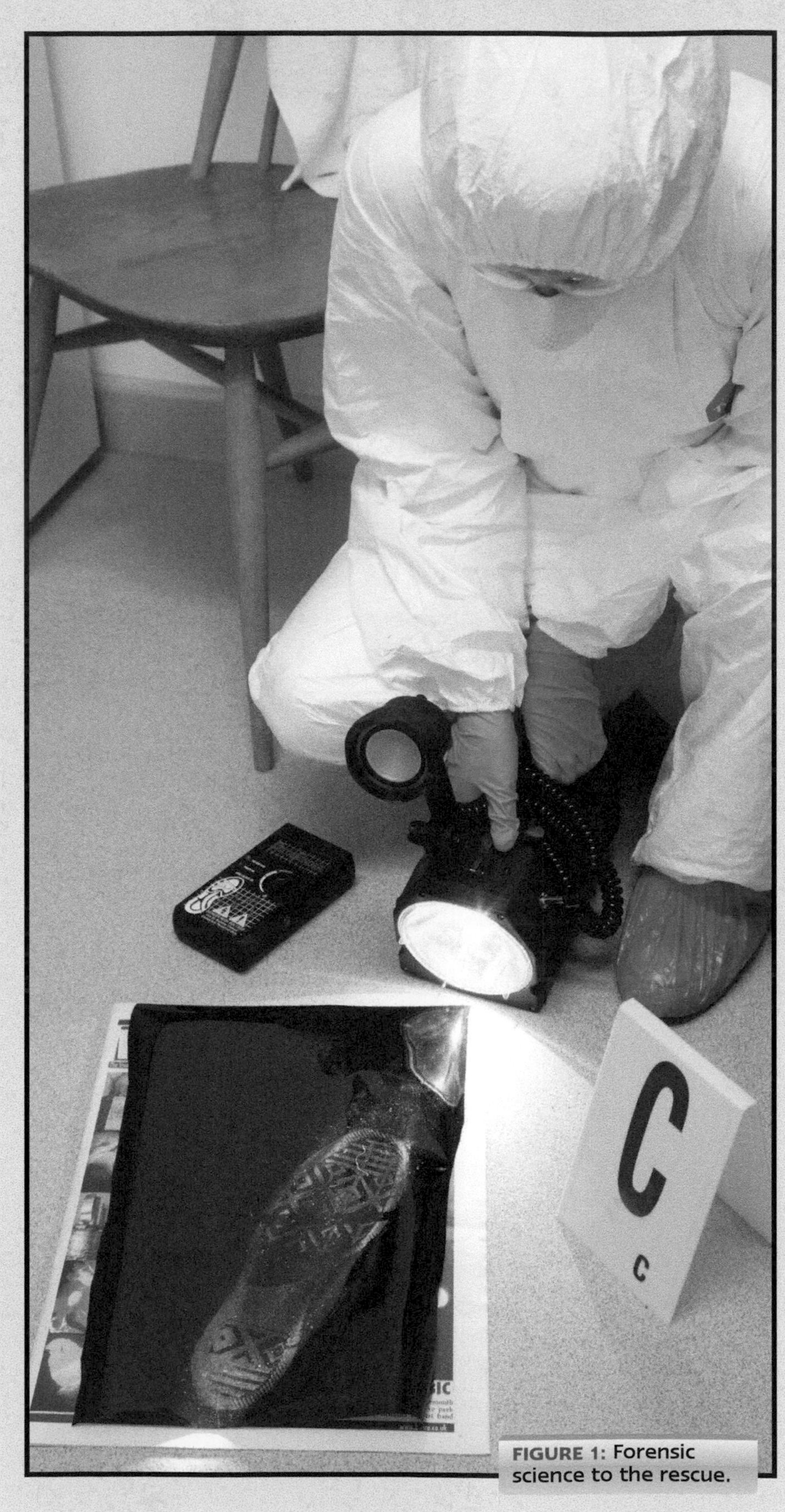

FIGURE 1: Forensic science to the rescue.

Assess Yourself

See details of chromatography on **pages 74** and **75**

1. Which scientific technique depends on the way that particles of ink move through paper?
2. Why would different solvents be needed for car paint and for marker pens?
3. How do you think the strength of forces between particles compares in car paint and marker pen ink?
4. How did the police prove that one particular pen was the one used to leave the messages?
5. What practical problem did the forensic scientists have to overcome when making a chromatogram of car paint?
6. How do you think that this problem was solved?
7. What other kinds of samples might have enabled the police to use chromatography?
8. Is it sufficient evidence to find that the paint on a car matches that at the scene of the crime?
9. Some colourless materials glow in ultraviolet light (U.V.). How might this help solve crimes?

ICT Activity

Using a digital camera, photograph the chromatograms produced in the chromatography experiment on pages 74 and 75.
Set up a database of ink colours with the corresponding ink spot patterns. Other students can now use this as reference material.

Maths Activity

Compile a table of all the pens and markers owned by students in your group. Display the data by colour of ink and whether it is permanent or washable. Decide how easily you could link one person to a particular message. What are the chances of your being right?

Level Booster

8 Your answers demonstrate extensive knowledge and understanding about how evidence can be evaluated and used to draw conclusions. You have shown a thorough grasp of an important application of Science.

7 Your answers show that you can synthesise information and identify key factors from a range of sources to use as evidence in solving crimes.

6 Your answers show that you can select and use sources of information and collect enough data to support your conclusions using chromatographic techniques.

5 Your answers show that you can select and use chromatographic techniques to obtain data in a systematic way.

4 Your answers show that you can use a fair test to answer the question of who carried out the crime.

Drinking water

BIG IDEAS

You are learning to:
- Recognise what makes water safe to drink
- Explain how to remove the salt from seawater
- Evaluate the choices available for making fresh water

Taking the salt out

One way to get drinking water from seawater is to take out the salt. This is called **desalination**. It uses a lot of energy to boil the water before the steam can be condensed as fresh water. The salt cannot boil, it is left behind when the steam boils off.

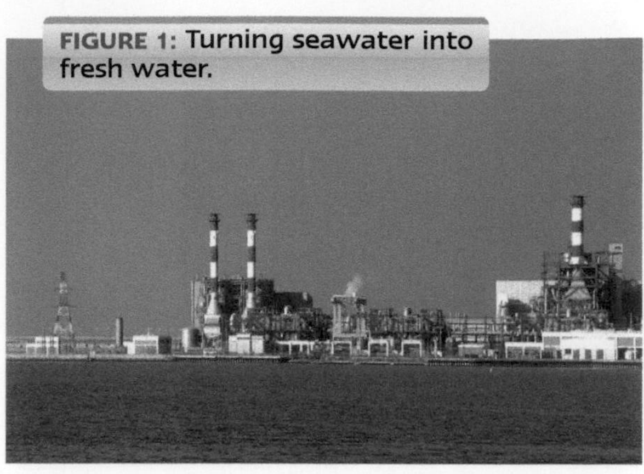

FIGURE 1: Turning seawater into fresh water.

Long ago people noticed that water does not always boil at the same temperature. At sea level water boils at 100 °C. This is the **boiling point** (b.p.) of water. If you climb a mountain, the b.p. of water changes. At the top of Mont Blanc in France, Europe's highest mountain, the b.p. falls to about 85 °C. The reason is that the air pressure is less on Mont Blanc and this affects the b.p. If you apply the same idea to distillation you can save both energy and money. The seawater is distilled in a large apparatus where the pressure inside is lower than normal. Water then boils more easily at a lower temperature. This method is used in hot countries which have low rainfall.

1. What do we mean by desalination?
2. Why is it expensive to turn seawater into fresh water?
3. How can the cost be reduced?

Freezing out the salt

When seawater freezes, the ice does not contain any salt. The remaining seawater becomes more salty. We say that the **salinity** increases. Some big desalination plants freeze salty water to obtain fresh water ice. When this ice is collected and melted it gives drinking water. The freezing method is used in some of the Baltic countries of Europe. The freezing method works best with water whose salinity is less than ordinary seawater. Seawater usually has about 35 000 parts of salt in every one million parts of water. In the Baltic this can be as low as 5000 parts of salt.

4. What do we mean by the salinity of seawater?
5. What happens to the salt when seawater freezes?
6. What fraction of normal salinity is found in the Baltic?

7. The density of salt water varies with the salinity. How could you use a drinking straw, weighted at one end to float upright, to compare salinities of water samples?

Did You Know...?

One idea for obtaining fresh water is to tow icebergs from the Arctic or the Antarctic to places lacking fresh water. When the icebergs arrive they could be left to melt. All you need to do is collect the fresh water. Nobody has found a way to do this yet.

FIGURE 2: Solid drinking water.

How Science Works

Extracting pure water from salty water

Imagine that you were given the task of turning salty water into pure water. How could you remove the salt to make it pure?

Your job is to come up with a plan to do this. You'll need to look at what you've got to achieve, think about what equipment to use and come up with some ideas. You then need to decide on the best idea to use, draw a diagram, label it and be ready to explain how it will work.

You can use various pieces of a student laboratory apparatus, but nothing as complex as a Liebig condenser.

You could heat the water to separate it from the salt but what could you then do with the vapour? Your job is to end up with pure water. Show your design to some other students to see what they think. See if you can get some ideas to improve it.

See what the teacher thinks: you may be allowed to try out your plan. If you are, you could make some notes about how well it worked.

Save the sailor!

Now imagine that a sailor travelling around the world runs out of drinking water while in the middle of the Pacific Ocean. The sailor only has some plastic bags, tubing, buckets and adhesive tape. Devise a way to make fresh water from the seawater, using the heat of the Sun.

HSW

Chromatography

BIG IDEAS

You are learning to:

- Devise a way to separate colours
- Explain what happens in chromatography

Separating colours

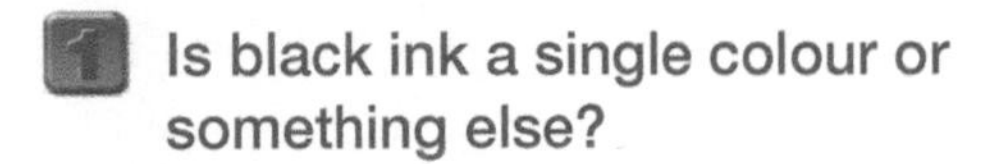

Black ink is not just a black colour mixed with water. Black ink is a mixture of colours. We can use filter paper and water to separate these colours. We call this **chromatography**. The pictures show what happens when the colours are separated. Drops of water are added to the middle of the paper where the ink spot was placed.

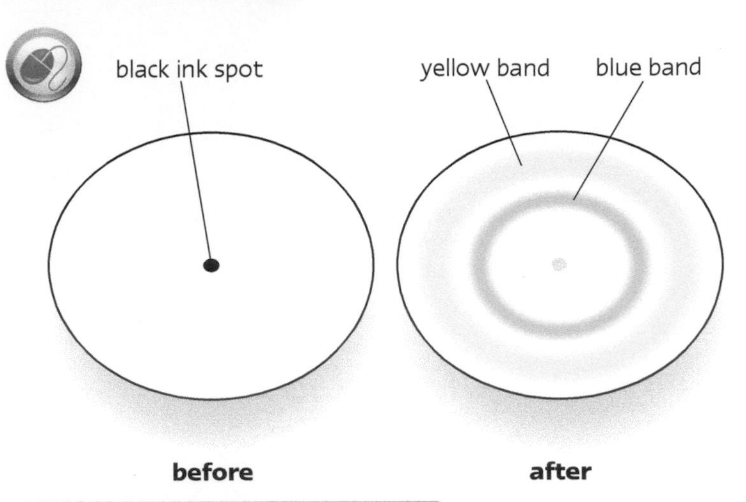

FIGURE 1: Separating ink colours.

1. Is black ink a single colour or something else?
2. What material do we need to separate the colours in ink?

Examples of chromatography

If you cut a section of paper, it can act as a wick. By dipping this wick into water, the liquid is drawn up through the ink and colour begins to separate all on its own. The resulting pattern of colours is a **chromatogram**. The second method is called **ascending** paper chromatography. The water soaks up from the base, carrying the colour spots with it. Some colours move faster than others. This is how we can separate the colours. You can use chromatography for colourless mixtures. At the end, you must develop the chromatogram by adding another material that makes the spots visible.

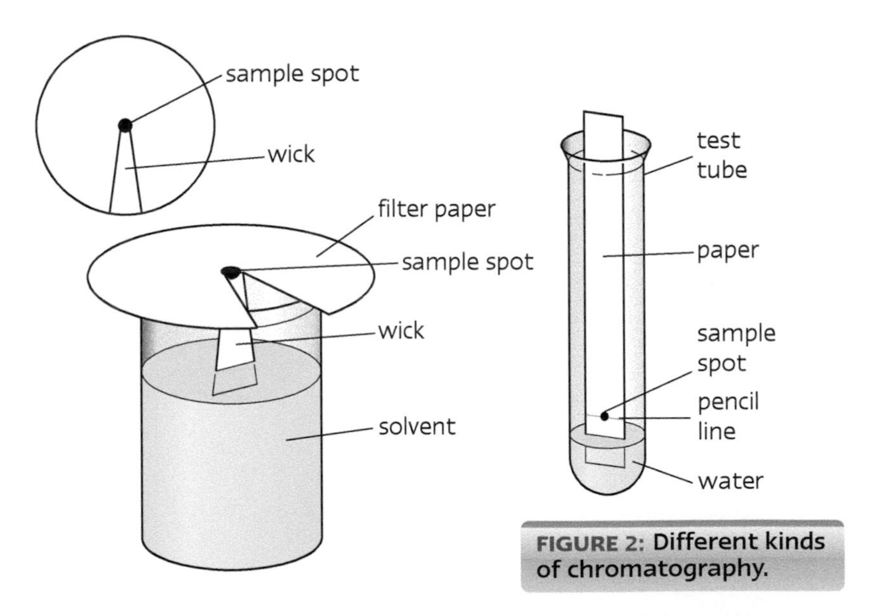

FIGURE 2: Different kinds of chromatography.

3. Why do lazy students choose the wick method?
4. What do we call the pattern of spots on the paper?
5. How might chromatography happen by accident?

Comparing colours

We can use paper chromatography to compare colours. The colours can be felt tip pens or ink or food dyes. If the colour mixtures are the same then they give identical chromatograms.

Method

1 Cut the paper strip to fit an empty beaker.

2 Draw a pencil line 2 centimetres above the base. Put four colour spots spaced out on this line.

3 Use a paper clip to make a paper cylinder (see diagram).

4 Place 1 cm depth of water in the beaker and put in the paper.

5 Wait until the water has soaked up nearly to the top.

6 Remove the paper and mark in pencil the **solvent front**, the highest level that the water reached.

7 Compare the colour spots. If the patterns are the same then the original colours were the same too.

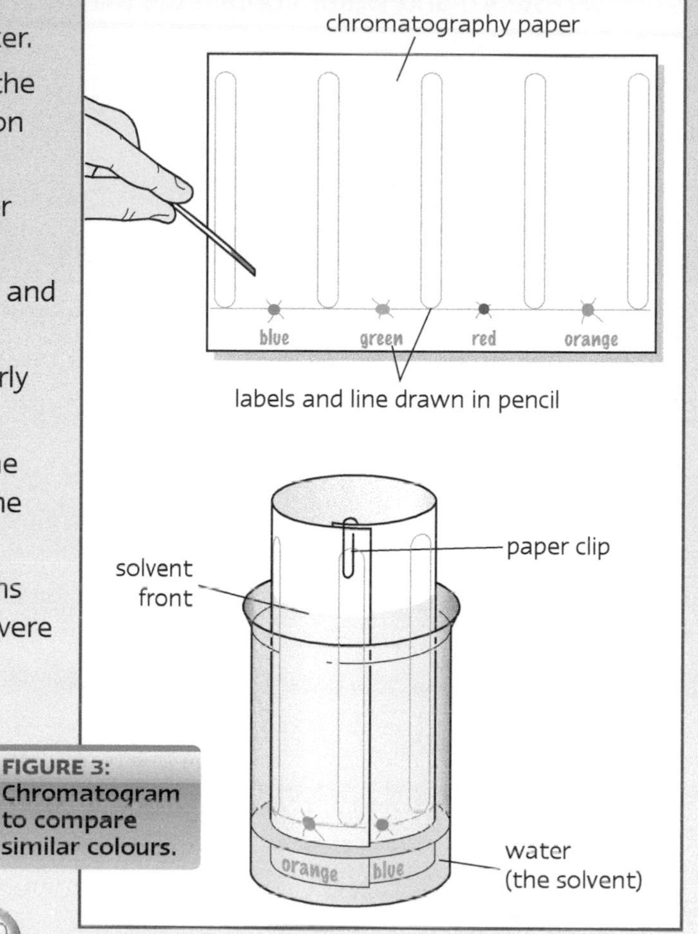

FIGURE 3: Chromatogram to compare similar colours.

8 The same colour moves the same distance along the paper.

Did You Know...?

If you use a fountain pen, the ink colour can run when the paper gets wet. Ink on paper or even ink stains on a white shirt can make their own chromatograms in the rain. It still makes a mess but the science is interesting.

Questions

1 Why is the line drawn in pencil and not ink?

2 Why is the starting line above the water level in the beaker?

3 If two brands of blue dye give identical spot patterns, what can you conclude?

4 What extra step is required when making a chromatogram of a colourless mixture?

5 Some dyes are invisible in normal light and only show up in ultraviolet light (U.V.). This is how security inks work. How could you use chromatography with a set of security inks?

Practice questions

1 For each of the following statements, write **T** if it is true or **F** if the statement is false.

- **a** Both salt and sugar dissolve in water.
- **b** Steam and water contain the same chemical material.
- **c** Breathing on a mirror shows that gases can condense.
- **d** We use chromatography to separate insoluble solids.

2 Copy the sentences a–d into your exercise book. Use the following words to help you answer the questions. The words may be used once, more than once, or not at all.

colours **condense** **evaporates** **insoluble** **soluble**

- **a** Water disappears from a puddle on the floor because it ________.
- **b** We use chromatography to separate ________.
- **c** Some paints are completely ________ in water.
- **d** Water vapour can ________ on the inside of a cold tube in a condenser.

3 Copy the columns below into your exercise book. Draw straight lines to join the word with the correct description.

Word	**Description**
a evaporation	very soluble
b alcohol and acetone	solution mining
c sugar	change of state
d salt is extracted by	solvents

4 Copy the columns below into your exercise book. Identify the correct separation technique for these mixtures.

Mixture	**Separation technique**
a sand from salt	a
b water from ink	b
c colours in ink	c
d salt from seawater	d

5 Rock salt is stirred with water and then filtered, the liquid is collected in a basin. Copy the sentences a–d into your exercise book and complete them by choosing from the words below:

residue **solute** **solution** **solvent**

- **a** The material left in the filter paper is ________
- **b** The salt is the ________
- **c** Water is the ________
- **d** In the basin is the ________

6 Use the information in the table to help you answer the following questions about a chromatography experiment.

Sample	Pattern on the chromatogram
Black ink A1	Yellow and blue
Black ink A2	No pattern seen
Black ink A3	Blue and yellow
Black ink A4	Yellow, green, dark blue

- **a** Which of the inks is insoluble?
- **b** Explain your answer to part (a).
- **c** Which inks were probably the same?
- **d** What extra information would you need to be certain of your answer to part (c)?

7 Put these steps in the correct order for the separation of pure water from sea water.

collection **condensation** **evaporation** **heating**

- a 1st step
- b 2nd step
- c 3rd step
- d 4th step

8
- a What happens to the boiling point of water at low pressures?
- b Which has the higher boiling point, alcohol or water?
- c What would be in the first distillate on distilling a bottle of wine?
- d Why is the boiling point of water about five degrees higher than normal when measured at the base of a deep coal mine?

9
- **a** Liquid A boils at a lower temperature than liquid B. In which liquid are the inter-particle forces stronger?
- **b** Which liquid's particles will be more common in the vapour above a mixture of liquids A and B at room temperature?
- **c** What information would a thermometer give you when distilling a mixture of A and B?
- **d** Name the process by which particles of B might spread in the air.

10
- **a** What physical property makes the solution mining of salt possible?
- **b** How could you investigate any variation in this property with temperature in the laboratory?
- **c** What would you predict would be the variation?
- **d** What implications does this have for the salinity of the oceans?

Learning Checklist

4

- ☆ I know how much sugar can dissolve. page 61
- ☆ I know that rock salt contains dirt. page 62
- ☆ I know that graffiti is hard to remove. page 64
- ☆ I know that gases can be cooled to give liquids. page 68

5

- ☆ I know how to make things dissolve. page 58
- ☆ I know how to measure the solubility of sugar. page 61
- ☆ I know that the impurities in rock salt are insoluble. page 62
- ☆ I know that special solvents are needed to remove graffiti. page 64
- ☆ I understand how distillation works. page 66
- ☆ I know how a condenser works. page 68

6

- ☆ I can identify the variables in solubility experiments. page 61
- ☆ I can interpret a solubility graph or tables of solubility data. page 61
- ☆ I can devise an experiment to purify rock salt. page 63
- ☆ I can test solvents to find the best one. page 65
- ☆ I can identify the changes of state in distillation. page 66
- ☆ I can explain the Liebig condenser. page 69

7

- ☆ I know some of the trends in solubility. page 61
- ☆ I can explain changes of state in terms of particles. page 63
- ☆ I can assess the benefits and problems of using particular solvents. page 65
- ☆ I can improve an experiment to purify seawater. page 67
- ☆ I know how to find the best method to purify liquid mixtures. page 67

8

- ☆ I know about the anomalous solubility of salt. page 62
- ☆ I can assess the hazards of using materials such as flammable solvents. page 64
- ☆ I can evaluate the evidence for different ways to separate mixtures of liquids and solids. page 72

Topic Quiz

1. What happens when sugar is stirred with water?
2. What do we call the liquid that dissolves things?
3. On the beach, how do we know that sand is insoluble?
4. Why can't we remove graffiti using soap and water?
5. What happens inside a condenser?
6. What is in rock salt?
7. Name the soluble part of rock salt.
8. What do we mean by desalination?
9. What happens to black ink in chromatography?
10. In chromatography, what is the name of the paper with the pattern of colour spots?
11. How could you be sure that two black inks are the same?

True or False?

If a statement is false then rewrite it so it is correct.

1. Sand is insoluble in water.
2. Water is not a solvent.
3. Fine sugar dissolves at the same speed as sugar lumps.
4. Separating dyes into colours is called chromatography.
5. Police scientists use chromatography to solve crimes.
6. In distillation there are two changes of state.
7. Distillation separates mixtures of liquids according to their colours.
8. Water and alcohol boil at fixed temperatures anywhere on the Earth.
9. The Liebig condenser uses a jacket of cold water to condense the vapour.
10. Solubility trends are shown clearly by solubility graphs.
11. Salt can be extracted by solution mining.
12. Brine is a solution of water and ink.

Literacy Activity

Write a letter from the owner of a salt works to a house owner warning them about solution mining. Make it clear that the house may collapse.

ICT Activity

Produce a flow scheme to show how salt can be extracted both by mining and by solution mining.

LEGO MODELS

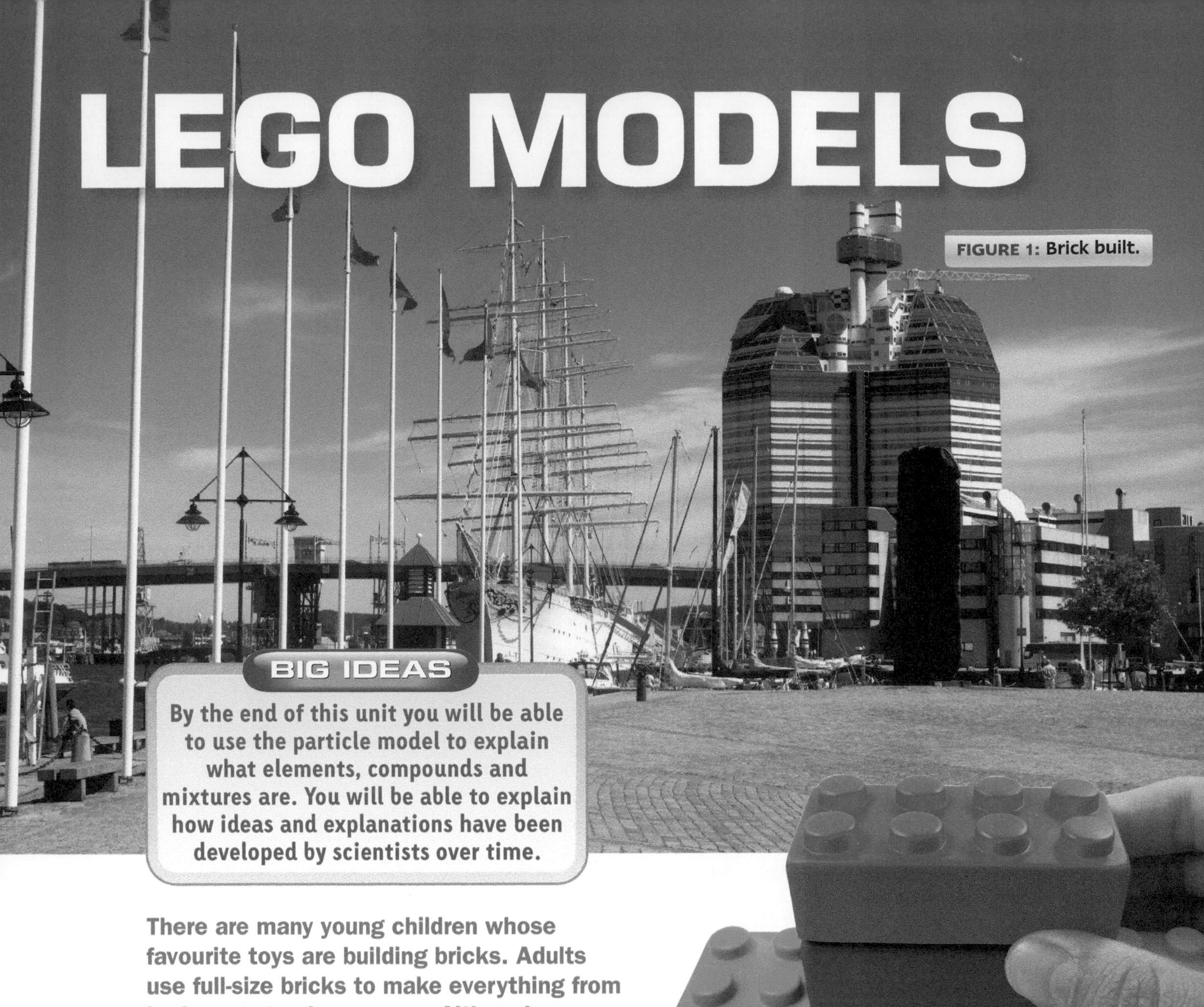

FIGURE 1: Brick built.

BIG IDEAS

By the end of this unit you will be able to use the particle model to explain what elements, compounds and mixtures are. You will be able to explain how ideas and explanations have been developed by scientists over time.

There are many young children whose favourite toys are building bricks. Adults use full-size bricks to make everything from barbecues to skyscrapers. Although very young children do not realise it, there are rules about how you can build with bricks.

You cannot choose parts of one brick, only single bricks. They can be fitted together only in particular ways. Children experiment and soon find how much larger things can be built from single units. On the Earth there are 92 elements that are found naturally. This is a bit like having 92 different kinds of building bricks. Elements are single materials, such as carbon and copper or oxygen and gold. The particles inside copper are called copper atoms. All of these copper atoms are the same. Gold is made of gold atoms.

FIGURE 2: The element gold.

Atoms, elements and compounds

FIGURE 3: Diamonds are made of carbon.

Sometimes separate atoms link into groups called molecules. Oxygen gas is like this. Two atoms are joined together. We write the molecule of oxygen as O_2 since the shorthand for oxygen is the letter O. The yellow element called sulphur has larger molecules of S_8. Eight sulphur atoms are joined together in a ring molecule.

When different elements combine we get new materials called compounds. Sulphur and oxygen can combine to form the compound SO_2. This molecule contains one sulphur atom joined to two atoms of oxygen. Once you understand how atoms combine, you start to understand just how chemistry works.

What do you know?

1. How many elements are found naturally?
2. What do we call the particles inside a piece of copper?
3. Which has more particles, an atom or a molecule?
4. A molecule of the gas ozone is written as O_3. What does this mean to you?
5. Which of these do you think are elements?

   a N_2 **b** NO_2 **c** NH_3 **d** H_2
6. In what ways are atoms like building bricks?
7. When the elements on the Earth combine to give compounds, how many are there likely to be: hundreds, thousands or millions?
8. How might the letters of the alphabet provide a model for how the elements combine?
9. Suggest **one** way in which atoms are probably different to building bricks.
10. Why do you think many scientists use models to understand and explain their ideas about particles?

Chemical alphabet

BIG IDEAS

You are learning to:

- Interpret some chemical symbols
- Describe just how many atoms are needed to make things

Alphabetical choices 

The English language has just 26 letters. There are more than 100 different **elements** but only 92 of them occur naturally on the Earth. The other elements have been made artificially. Since many of the names of the elements are complicated, we need a chemical shorthand for each of them. This is called a **chemical symbol**. There are not enough letters to go around. The rule for symbols is simple. Elements with a single letter have a capital letter. Elements whose symbols have two letters have just the first letter as a capital.

Exam Tip!

You do not have to learn the names and symbols of all the elements. You will just learn the common symbols as you use them.

Name	Symbol	Old name of the element
Hydrogen	H	–
Oxygen	O	–
Carbon	C	–
Cobalt	Co	–
Iron	Fe	Ferrum
Gold	Au	Aurum
Copper	Cu	Cuprum
Magnesium	Mg	–
Nitrogen	N	Azote
Sulphur	S	Brimstone
Calcium	Ca	–
Chlorine	Cl	–

1. What is the chemical symbol for carbon?
2. Are CO and Co the same thing?
3. **a** Name **two** elements whose symbols do not match their names.
   b What does the formula H_2O show us about water?
   c How many different elements are there in limestone, $CaCO_3$?
   d What are the names of all the elements in glucoise, $C_6H_{12}O_6$?

Fame at last

The names and symbols of some elements derive from famous people, places or other languages.

4. Give **one** use of the element whose symbol is Sr.
5. Scrapyards often advertise for **non-ferrous** metals. What does this mean?
6. The fictional character Goldfinger had an unusual first name, it was Auric. Explain why this was a good name.

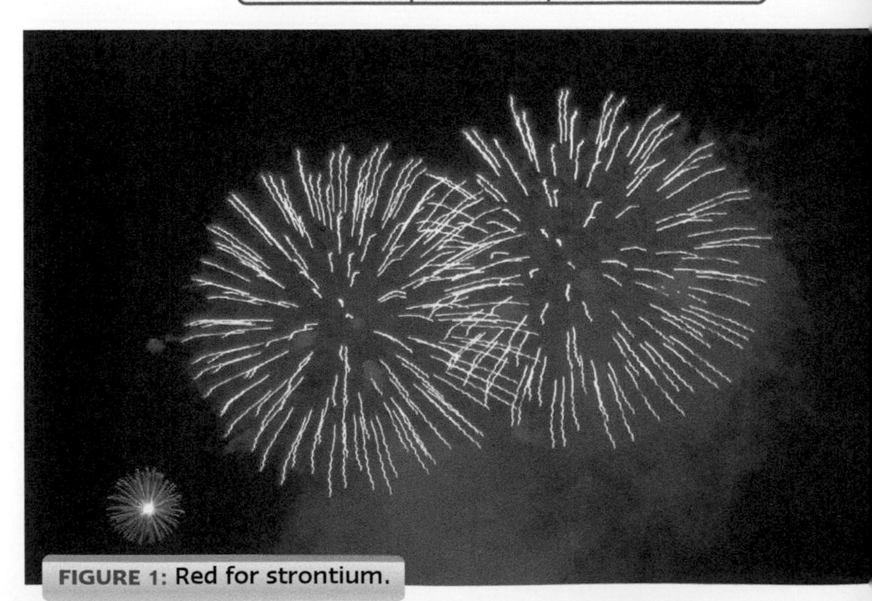

FIGURE 1: Red for strontium.

Element	Symbol	History
Strontium	Sr	Used to give a red colour to fireworks. Named after a town in Scotland called Strontian
Nickel	Ni	Named devil's copper or kupfernickel in German
Francium	Fr	Named after France
Einsteinium	Es	Named after Albert Einstein

Did You Know...?

The element mercury, the liquid metal used in thermometers, was named after Mercury, the Roman messenger of the gods. Mercury moved fast to deliver messages and mercury metal runs about when spilled. The symbol is Hg which comes from the Greek word meaning liquid silver.

Coin challenge

A one-penny piece is covered in copper metal and is about 2 cm across. Imagine that there is a single line of copper atoms stretching right across the coin. Many secondary schools have 1000 students. We can calculate that there are about 75 million copper atoms making up this line. If each student counted one copper atom every second, then 1000 students would need 24 hours to count all of them.

FIGURE 2: How many copper atoms can you get for one penny?

Think of a single drop of water falling from a dripping tap. If you could magnify this single drop until it was the size of the Earth, you would then be able to see the atoms inside. Each atom would be about the size of an apple.

FIGURE 3: This single drop of water contains millions of atoms.

7 Which number is larger, the 75 million copper atoms or the number of people living in Britain?

8 How many atoms could one student count in 24 hours at a rate of one atom each second?

9 Helium gas is made of single atoms. Even if a balloon is perfectly sealed, the gas escapes. Explain how.

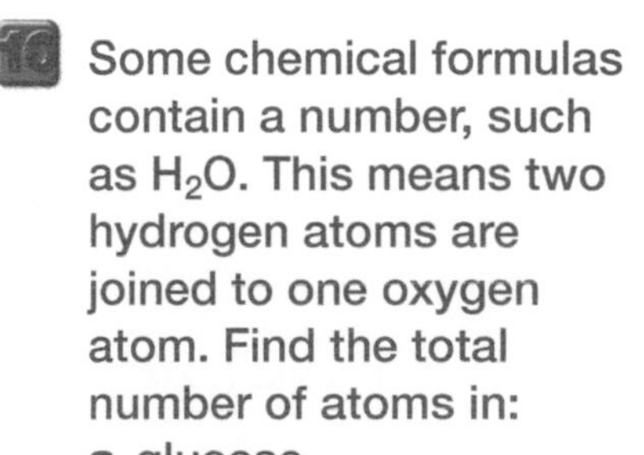

10 Some chemical formulas contain a number, such as H_2O. This means two hydrogen atoms are joined to one oxygen atom. Find the total number of atoms in:

a glucose
b limestone
c sucrose, $C_{12}H_{22}O_{11}$.

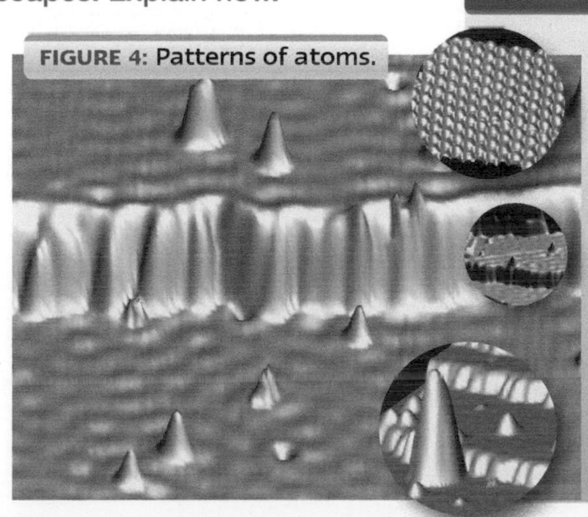

FIGURE 4: Patterns of atoms.

Did You Know...?

Ordinary **microscopes** that use light are no use for seeing atoms. The atoms are much too small to show up. Special electron microscopes are able to show images of atoms. In metals, we find the atoms are all lined up in rows in a regular pattern.

Getting sorted

BIG IDEAS

You are learning to:
- Recognise metals and non-metals
- Understand how elements are arranged in patterns in the Periodic Table HSW
- Recognise why the table of elements is useful

5

Metals and non-metals

Dividing the elements according to the states of matter is not very useful. It is much better to divide them into metals and **non-metals**. This covers almost all of the elements, with just a few exceptions left over. Each kind of element has its own special **properties**.

When you check the properties of the elements there are some exceptions. The non-metal graphite, one form of carbon, conducts both heat and electricity. Graphite is used to make ordinary pencils, it is soft enough to mark paper.

FIGURE 1: Metals and non-metals. A roll of aluminium foil, sometimes called tin foil, and a heap of sulphur.

1. What is unusual about the element Hg?
2. Non-metals are supposed to break easily but oxygen does not. Explain why.
3. Metals are said to be shiny but old coins are dull. Explain why.

Properties of metals	Properties of non-metals
Solid at room temperature (not Hg)	Solids, liquids or gases at room temperature
Shiny if polished	Not shiny
Conduct electricity	Do not conduct electricity (except graphite, a form of carbon)
Conduct heat	Do not conduct heat (except graphite)
Bend without breaking (**malleable**)	Break easily if solid (**brittle**)
Can stretch into wires (**ductile**)	Cannot stretch easily

Did You Know...?

There are millions of tonnes of gold dissolved in the sea. If you could recover this gold you would be very rich. Unfortunately, the cost of extracting the gold is more than the gold is worth.

... atomic number ... brittle ... ductile ... group

The Periodic Table

The **Periodic Table** displays the symbols of the elements in a special order. The table was first developed in the nineteenth century, following careful observations of the elements. We could show the symbols in different ways. They could be in alphabetical order, or arranged by colour or using some other property. The Periodic Table is much more useful. Every element has its own number, the **atomic number**.

metals

non-metals

				H 1													He 2
Li 3	Be 4											B 5	C 6	N 7	O 8	F 9	Ne 10
Na 11	Mg 12											Al 13	Si 14	P 15	S 16	Cl 17	Ar 18
K 19	Ca 20	Sc 21	Ti 22	V 23	Cr 24	Mn 25	Fe 26	Co 27	Ni 28	Cu 29	Zn 30	Ga 31	Ge 32	As 33	Se 34	Br 35	Kr 36
Rb 37	Sr 38	Y 39	Zr 40	Nb 41	Mo 42	Tc 43	Ru 44	Rh 45	Pd 46	Ag 47	Cd 48	In 49	Sn 50	Sb 51	Te 52	I 53	Xe 54
Cs 55	Ba 56		Hf 72	Ta 73	W 74	Re 75	Os 76	Ir 77	Pt 78	Au 79	Hg 80	Tl 81	Pb 82	Bi 83	Po 84	At 85	Rn 86
Fr 87	Ra 88		Rf 104	Ha 105	Sg 106	Ns 107	Hs 108	Mt 109									

La 57	Ce 58	Pr 59	Nd 60	Pm 61	Sm 62	Eu 63	Gd 64	Tb 65	Dy 66	Ho 67	Er 68	Tm 69	Yb 70	Lu 71
Ac 89	Th 90	Pa 91	U 92	Np 93	Pu 94	Am 95	Cm 96	Bk 97	Cf 98	Es 99	Fm 100	Md 101	No 102	Lw 103

FIGURE 2: Displaying the elements.

The table is in the order of the atomic numbers. This produces some interesting patterns in the table. Metals are all on one side, non-metals on the opposite side. The gases are all together too. The vertical columns are called **groups**. Elements in the same group have very similar chemical properties.

How Science Works

Scientists find it very useful to arrange things in ways that show patterns. The pattern in the data often leads to new discoveries. Errors may also be easier to spot when there is a clear pattern.

HSW

4 Why do we use symbols and not names in the Periodic Table?

5 On which side of the table are the metals?

6 What is special about elements placed in the same group?

Liquids and metalloids

At room temperature (25 °C) there are only two liquid elements. They are the metal mercury and the non-metal bromine. Other elements do melt to liquids if you heat them enough. The semi-conductor elements are all found together in the table. Look for Si, Ge, Ga and As.

7 Which unusual non-metal is found at the top of the same group as silicon?

8 Which elements are used to make the material gallium arsenide that is used to make high-speed computers?

9 Which balloon gas can you find in the last group?

10 Locate the following in the Periodic Table.
a **Three** magnetic metals: iron, cobalt and nickel. Give their symbols.
b The unreactive gases helium, neon and argon. Give their symbols.
c The expensive metals gold, silver and platinum. Give their symbols.

FIGURE 3: A liquid non-metal.

HSW Atomic Thoughts

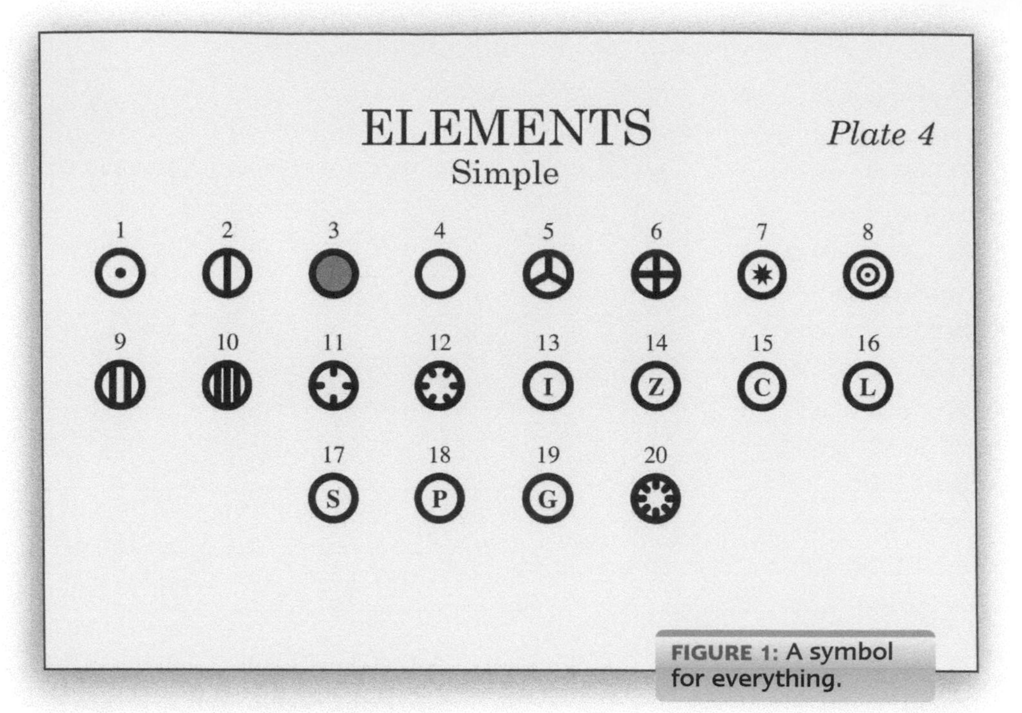

FIGURE 1: A symbol for everything.

The Prof does it again

Last week here in Manchester our famous Professor, Mr Dalton, announced his new theory. He paid tribute to earlier scientists whose ideas he had developed further. The Greek philosophers Leucippos and his student Demokritos were known as the Greek Atomists, around 500 BC. Professor Dalton has developed their ideas so that we can now understand just what chemistry is all about. He is writing a book about his new ideas, to be called a 'New System of Chemical Philosophy'. When asked by our reporter to explain his ideas in a few words, Professor Dalton said it was all quite simple.

1807

- Everything in the world is made of tiny particles.
- The particles are called atoms and cannot be split up.
- Atoms cannot be created or destroyed.
- Each element has its own special kind of atom and they are all the same as each other.
- Different elements have atoms with different masses.
- When elements combine, it is always between small numbers of the atoms. They form compound-atoms, such as water.

Professor Dalton has used his ideas to work out the formulae of many compound-atoms.

Name	Modern name	Formula using Dalton's symbols	Modern formula
Azote	Nitrogen		N_2
Firedamp	Methane		CH_4
Carbonic acid gas	Carbon dioxide		CO_2

Assess Yourself

1 How is it possible to get salt out of rocks just by using water?
2 What evidence from inside kettles suggests that some rocks dissolve?
3 What kind of solvent might remove a ballpoint pen mark on a shirt?
4 Why is it possible to separate mixtures of liquids by boiling?
5 What sorts of information can you get from the Periodic Table?
6 Why is it generally easier to separate mixtures rather than compounds?
7 What assumptions did Dalton make about atoms?
8 What can the formula of a compound such as H_2SO_4 tell us?

ICT Activity

Choose any **ten** elements and design your own pictorial symbols, in the style of Dalton. Combine some of them to display compound-atoms (molecules).

History Activity

Research the history of the atomic theory by finding out about John Dalton (1766–1844), and **either** Gay-Lussac (1778–1850) **or** Avogadro (1776–1856).

Level Booster

8 You have displayed extensive knowledge and understanding relating to atomic theory and have confidently used chemical formulae to explain concepts.

7 Your answers show you can effectively use information from a range of sources to explain complex ideas, such as the atomic theory.

6 Your answers show a good use of sources of information to explain the particle theory of matter and its applications.

5 Your answers show that you can use data systematically to provide an explanation of basic particle theory and the workings of separation techniques.

4 Your answers show that you understand how to use a fair test to explain something and that you can use some scientific words correctly.

All mixed up

BIG IDEAS

You are learning to:

- Explain what we mean by a mixture
- Clarify the properties of mixtures
- Separate mixtures effectively

Mixed fruit

If you put apples and oranges together, it makes a **mixture**. It is easy to separate mixtures. In this case, just picking up all the apples will separate the mixture. You can have mixtures of elements too.

FIGURE 1: A mixture.

1. Do you think you can separate iron powder mixed with lead powder? Explain why this is probably possible.
2. Is there a chemical change when you just mix apples and oranges together?

More mixtures

There are only three metals that are attracted by a **magnet**. These metals are iron, cobalt and nickel. We say these metals are **magnetic**. The particles of each metal are attracted by a magnetic field. This gives us one way to separate these elements from mixtures with other materials. The table gives a summary of the important properties of mixtures.

	Rule for mixtures	Example
1	Mixtures can be separated by physical methods.	Use a magnet to separate iron from non-magnetic carbon.
2	Mixtures just have the properties of the things in the mixture.	Iron fizzes with acids but sulphur does not. In a mixture of iron and sulphur, only the iron fizzes.
3	Mixtures of elements can be made using different amounts of each one.	1 part iron with 10 parts sulphur is a mixture, and so is 10 parts iron and 1 part sulphur.
4	No chemical change occurs when making mixtures.	Mixing iron and sulphur powders causes no temperature change, no light, and no new material is produced.

3. Name **two** magnetic elements other than Fe.
4. How might a thermometer help you to decide if there was any chemical change on mixing iron and charcoal?
5. Would all of a mixture containing nickel and lead powders be magnetic?

FIGURE 2: Iron is magnetic.

... *magnet* ... *magnetic*

FIGURE 3: Air is a mixture.

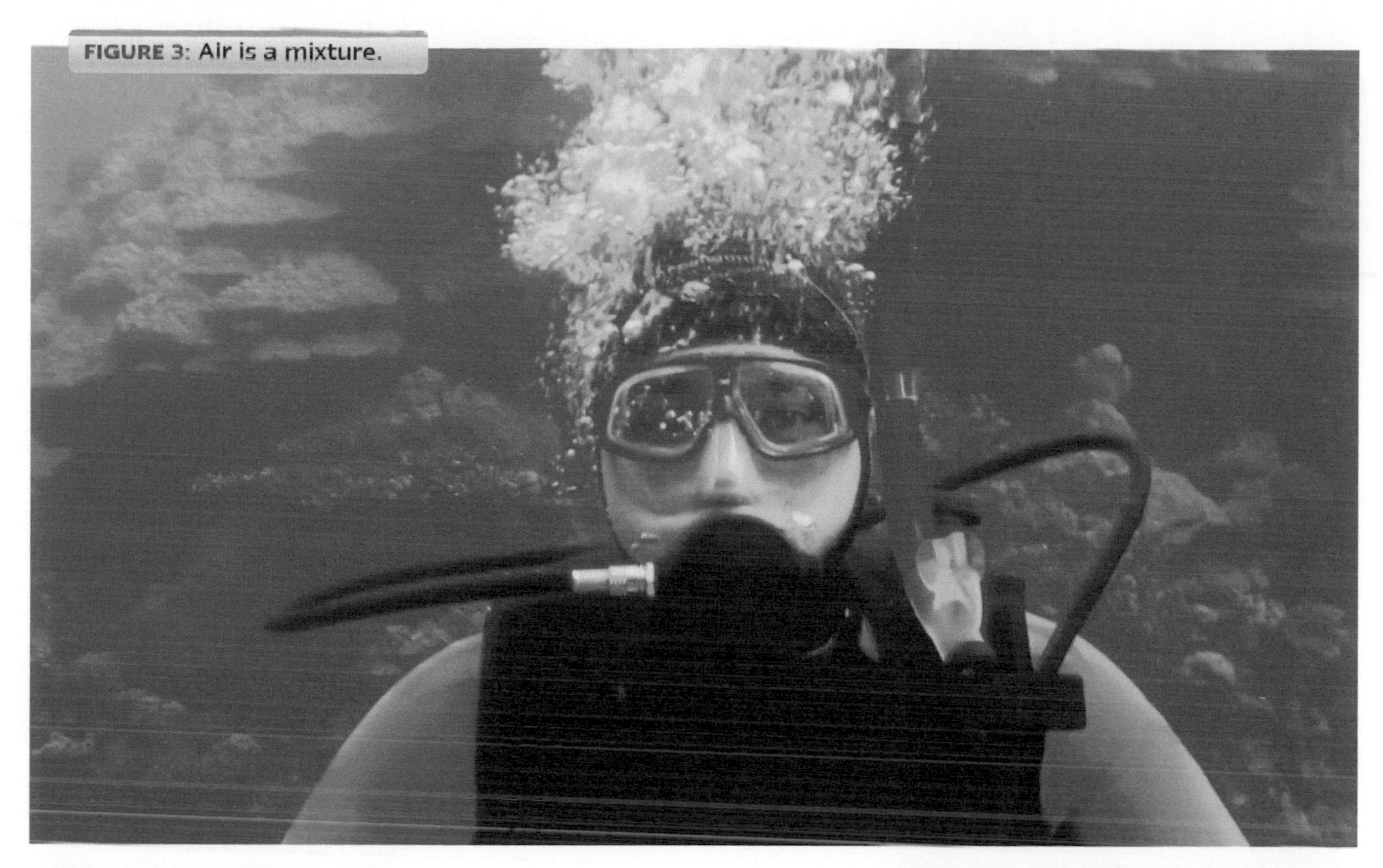

Up in the air

Just two gases make up 99% of the air we breathe. Most of the air is the element called nitrogen. Almost all of the rest is oxygen, the gas we need to breathe. The particles in both nitrogen and oxygen are molecules with two atoms joined together. We write this as N_2 for nitrogen and O_2 for oxygen. It is not obvious how you could separate the gases of the air. The gases must first be cooled down until they turn to liquid. Liquid nitrogen and oxygen boil at different temperatures. We can separate the two main gases in liquid air by boiling.

6 Newborn babies sometimes need to breathe pure oxygen. Where might this oxygen be obtained?

7 Which **one** of the following is not a physical method of separation?
a filtration **b** boiling
c combustion **d** freezing

8 Which is the second most common gas in the air?

9 Two other gases in the air are called argon (Ar) and carbon dioxide (CO_2). Give **two** similarities between these materials and **one** way in which they are different.

Did You Know...?

Gold dust can be separated from a mixture of sand and dirt by panning. The gold particles are much heavier, more dense and sink to the bottom. Rinsing with water separates the lighter materials, which can be washed away leaving the valuable gold behind.

FIGURE 4: Panning for gold.

What are compounds?

BIG IDEAS

You are learning to:
- Explain what we mean by compounds
- Recognise how compounds form
- Make compounds from elements 

4 Elements into compounds

Gold can make people do crazy things. When gold was first found in California many years ago, there was a gold rush. People rushed there to find gold and make themselves rich. Even the tiniest speck of gold dust still contains millions of gold particles, all the same as each other. Unfortunately, some other materials can look the same as gold. Fools' gold has fooled many people. It looks similar to the real thing. Fools' gold is a **compound**. The elements iron and sulphur have combined to give a new material. The scientific name for it is iron sulphide.

FIGURE 1: Fools' gold.

1. How did fools' gold get its name?

2. Which elements combine to give fools' gold?

5 Zinc and sulphur

Zinc is the useful metal that we put on iron to stop the iron going rusty. Zinc forms a compound with sulphur too. It gives zinc sulphide, known as the **mineral** zinc blende or sphalerite. In zinc sulphide, zinc particles have combined chemically with sulphur particles. The new particles of the compound zinc sulphide have completely new properties. The chemical reaction between the two elements can be violent, like a bad firework. Iron and zinc can also combine with oxygen in the air. They both form compounds known as **oxides**.

Did You Know...?

A sailing ship once returned from America filled with enough gold to make the crew very rich. There was just one problem when they tried to sell the gold. It was all fools' gold.

3. Which chemical compound do we find in zinc blende?
4. Name the compound formed when iron reacts with oxygen.
5. Why is it a bad idea to heat zinc and sulphur together?

FIGURE 2: A zinc compound – zinc blende.

... compound ... mineral

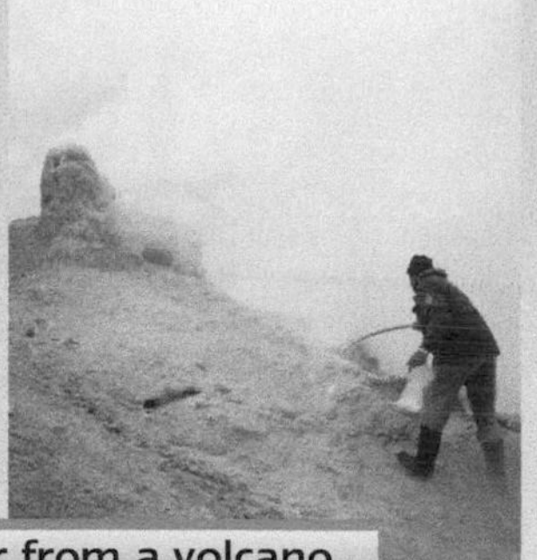

FIGURE 3: Sulphur from a volcano.

You can make compounds quite easily by heating **mixtures** together. The names of the compounds tell us which elements have combined.

Element 1	Element 2	Compound formed
Iron	Oxygen	Iron oxide
Magnesium	Oxygen	Magnesium oxide
Copper	Sulphur	Copper sulphide

Method:

1 Light a Bunsen burner and use a blue flame.

2 Using a metal spatula, drop a few iron filings into the top of the flame.

3 Using tongs, burn a 1 cm piece of magnesium in the flame.

Safety: Do not look directly at burning magnesium.

4 Using a test-tube, heat a mixture of yellow sulphur with brown copper powder. Look for any changes.

5 Copy the following table and complete it.

Number	Name of compound	Any changes you saw
1	*Iron oxide*	
2	*Magnesium oxide*	
3	*Copper sulphide*	

Questions

1 Which of the compounds was white in colour?

2 Which elements caused sparks when they combined?

3 Why is it difficult to describe the appearance of the compound formed when you burn carbon in oxygen?

4 Gold is an unreactive metal but fools' gold is the compound iron sulphide. What kind of chemical investigation might distinguish between them?

Understanding equations

BIG IDEAS

You are learning to:

- Select words to describe changes that we see (HSW)
- Write word equations
- Use symbols in equations

Burning hot 

The air contains **oxygen** gas. Oxygen is an active gas. It combines with other elements. In a barbecue the carbon burns away. In chemistry we write **word equations** to show what happens.

carbon + oxygen ⟶ carbon dioxide

The carbon turns into a new compound called an **oxide**.

FIGURE 1: Burning carbon.

1. Which reactant comes from the air?
2. What is formed when carbon burns?

A closer look

Oxygen and sulphur have similar **chemical reactions**. Sulphur burns in oxygen with a blue flame. It forms the oxide of sulphur called sulphur dioxide. The letters 'di' mean two, so dioxides contain two oxygen atoms.

There is another way to show this reaction without using any words. We can use **symbols** for each element.

Using the key in Figure 3 you can see what these equations mean

a 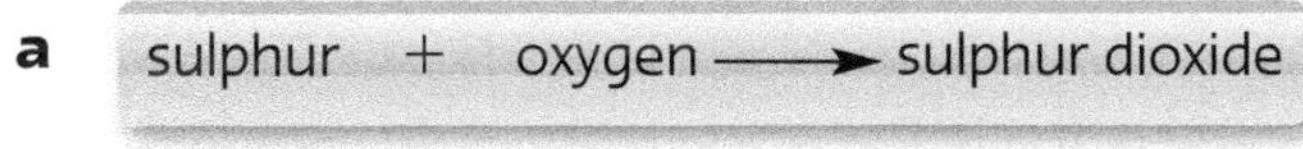

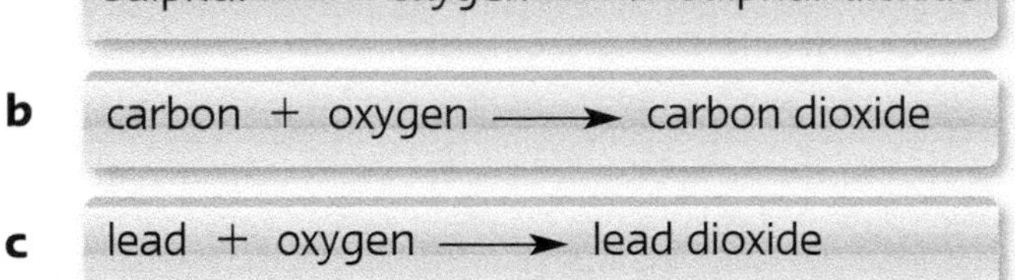

FIGURE 2: Sulphur burning in a bottle of oxygen.

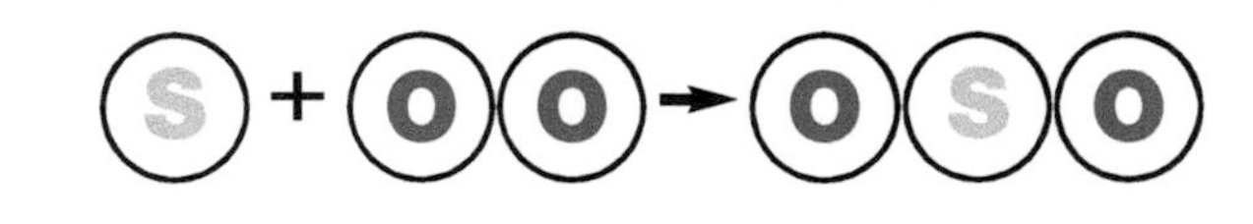

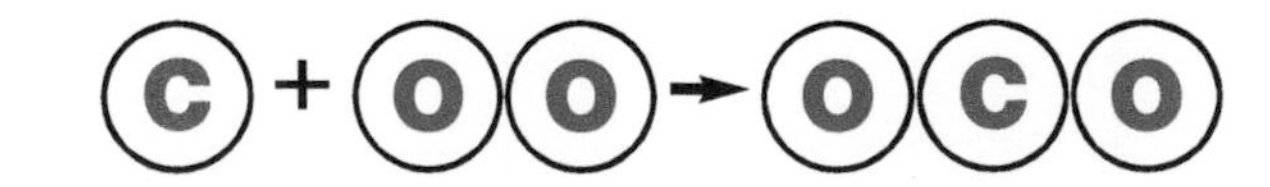

Pb + O O → O Pb O

C carbon
O oxygen
Pb lead
S sulphur

FIGURE 3: Another kind of equation.

If we have more symbols we can write other equations.

3. What is formed when elements combine with oxygen?
4. Complete the word equation:
   Lead + ______ ⟶ lead oxide
5. Write a symbol equation for the reaction in question 4.

Chemical symbols

The Periodic Table displays the symbols of all the known elements. Instead of using our own symbols, we can use chemical symbols from the table. The previous equations now look like this:

a $S + O_2 \longrightarrow SO_2$ The **formula** of sulphur dioxide is SO_2

b $C + O_2 \longrightarrow CO_2$

c $Pb + O_2 \longrightarrow PbO_2$

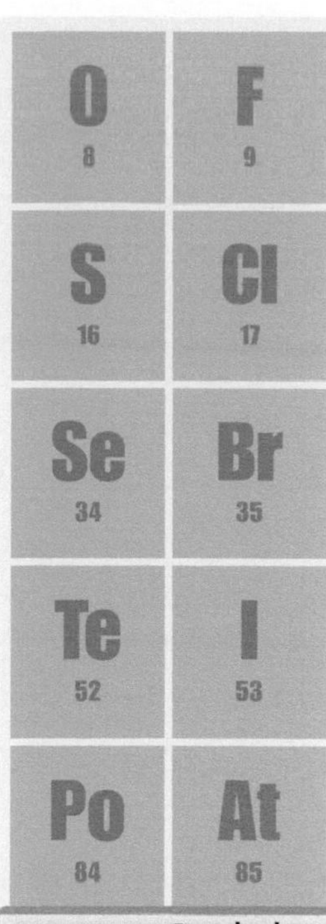

FIGURE 4: Symbols in the Periodic Table.

The same symbols are used by scientists all over the world. This makes it easier to understand reactions written by scientists in other languages.

The sand on the beach is another oxide called silicon dioxide. It is the most common mineral in the Earth's crust.

6 Write a word equation for silicon reacting with oxygen.

7 Write a symbol equation using chemical symbols for the same reaction. Silicon is Si.

8 Write down the formula of:
a sulphur trioxide **b** carbon monoxide.

9 Suggest possible names for these compounds:
a NO **b** FeO **c** ZnO **d** FeS **e** NI_3

Did You Know...?

Some types of silicon dioxide are used in jewellery. For example, amethyst, citrine and jasper.

FIGURE 5: Amethyst.

Chemical reactions

The important thing to remember about chemical reactions is that they are a way of rearranging atoms. Look at the first equation above. On the left hand side there is one atom of sulphur and two of oxygen. On the right hand side there is one of sulphur and two of oxygen.

This is always true in a chemical equation. You should always check to make sure this is true when you write down an equation.

10 Look at your answer to question 7 and see if you have the same number of atoms of silicon on both sides and the same number of atoms of oxygen on both sides.

If you did not have the same number of atoms of a particular element on both sides, what very odd thing would this mean was happening?

Did You Know...?

Insects send each other chemical signals. A female silk moth can attract a mate from several kilometres away. The female releases the chemical compound called bombykol. There is real chemistry between them.

Combining elements

BIG IDEAS

You are learning to:
- Link play bricks to the idea of elements combining (HSW)
- Explain how we use mass
- Discuss the simple ratios found in chemical compounds

Toy compounds 

When young children play with bricks, they join them together. Figure 1 shows two ways to join bricks. One red brick can be stuck to one yellow or two yellow bricks. We say that the **ratio** of yellow to red is one to one (1:1) or two to one (2:1). Elements can join in a similar way.

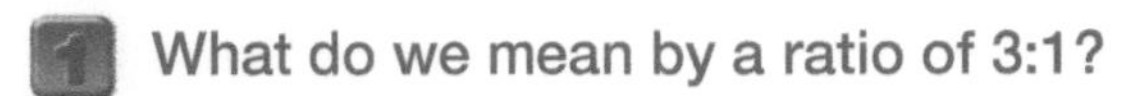

1. What do we mean by a ratio of 3:1?

2. If four yellow bricks are joined to one red, what is the ratio?

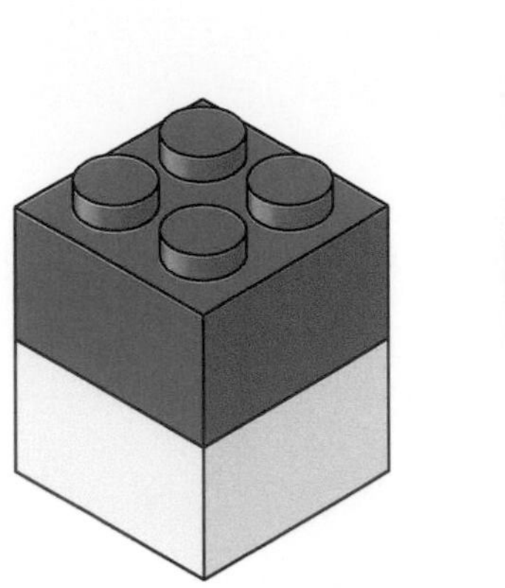

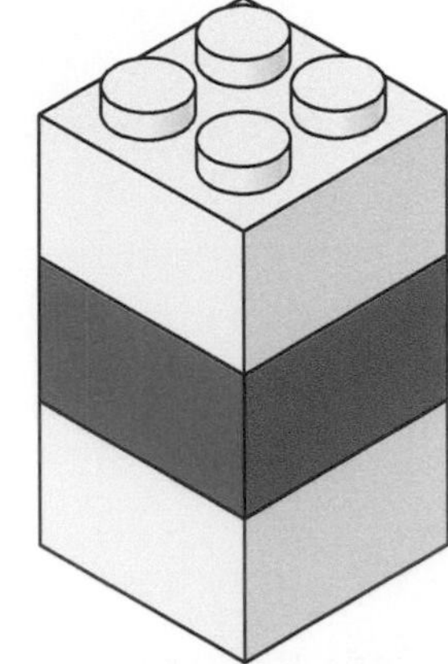

FIGURE 1: Combining toy bricks.

The importance of weighing

Lead is a dense metal and oxygen is an invisible gas. When these two elements combine they make lead **oxide**. It is more complicated than this because there are three different lead oxides. The chemical symbol for lead is Pb.

Oxide colour	Chemical formula	Ratio lead:oxygen atoms
Yellow	PbO	1:1
Brown	PbO_2	1:2
Red	Pb_3O_4	3:4

When elements combine, the numbers of particles of each are always in a small whole number ratio. It is not likely that you will find a ratio such as 1:2:7 or even 25:63. This is known as the Law of Multiple Proportions; the ratios are always simple ones.

The ratio is found from measurement of the masses of the elements that combine together in a compound.

3. What is the chemical symbol for lead?
4. What is the ratio of lead to oxygen in the brown lead oxide?

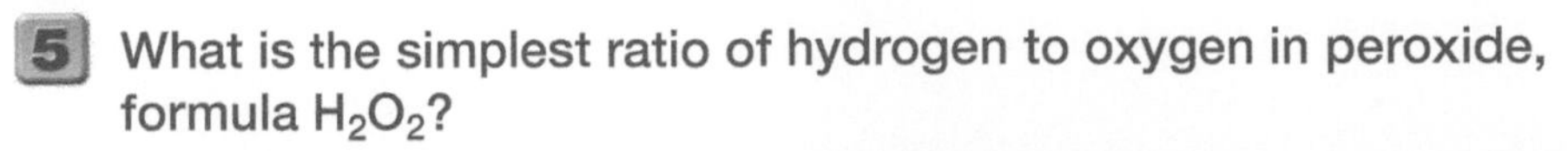

5. What is the simplest ratio of hydrogen to oxygen in peroxide, formula H_2O_2?

FIGURE 2: The oxides of lead metal.

Using data 

On a clean polished surface, lead is as shiny as cooking foil. On roofs, the lead loses its shine as the surface reacts with air and moisture. We can use careful weighing to find the masses of lead and oxygen that combine together.

Mass of oxygen (g)	Mass of lead (g)	Colour of oxide
1.6	10	Brown
0.8	10	Yellow

The masses of lead are the same here. The masses of oxygen are in the ratio of 0.8:1.6 or 1:2.

The two elements oxygen and nitrogen also combine to give more than one oxide.

FIGURE 3: Lead loses its shine.

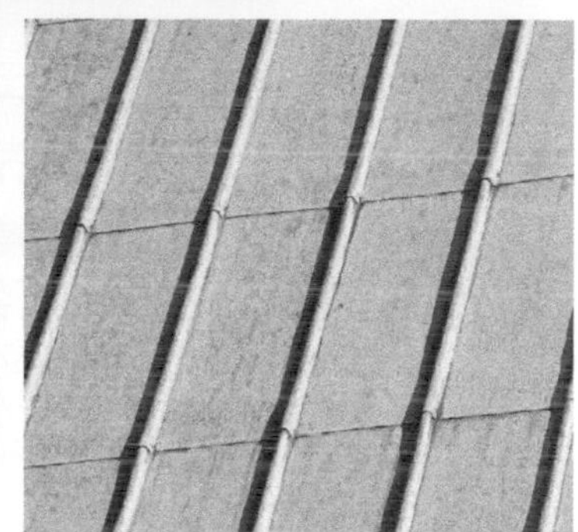

Mass of oxygen (g)	Mass of nitrogen (g)
57	100
114	100
171	100
228	100

By dividing all the oxygen values by the smallest one (57) we get the following ratio of oxygen combined with the same amount of nitrogen, 1:2:3:4

6 What would the ratio be for 285 g oxygen with the same mass of nitrogen?

7 What is the simplest ratio of the elements in glucose $C_6H_{12}O_6$?

8 If you heat lead oxides, they turn into different oxides. How could you tell that this has happened?

9 When you heat magnesium in air, it burns to leave a white ash. The reaction is violent, so we use small samples such as 0.6 mg magnesium. How would you expect the mass to change and how precise should the mass readings be?

How Science Works

Scientists need to exchange their results to check if they are right. This is why they must use the same units. On one space flight, things went wrong because the two construction teams building the spacecraft had used different units.

Practice questions

1 For each of the following statements write **T** if the statement is true or **F** if it is false.

a Mixtures and compounds are the same.

b It is easy to separate mixtures.

c Most of the elements in the Periodic Table are metals.

d One state of matter is called a solid.

2 Copy the sentences a–d into your excercise book. Use the following words to help you answer the questions. The words can be used once, more than once, or not at all.

conduct **insulator** **mercury** **non-metals**
sodium **symbols**

a All metals can ______ electricity.

b There is a liquid metal called ______.

c Chemical ______ are a useful shorthand for the elements.

d Most elements are either metals or ______.

3 Copy the columns below into your exercise book. Draw straight lines to join the word with the correct description.

	Word	Description
a	compound	chemical shorthand
b	element	two or more elements not combined
c	symbol	simple material
d	mixture	two or more elements that are combined

4 Copy the columns below into your exercise book. Choose the correct symbol or formula to match one of the names from this list.

diamond **nitrogen** **oxygen** **water**

	Symbol or formula		Name
a	O_2	a	
b	C	b	
c	H_2O	c	
d	N_2	d	

5 Copy the table below into your exercise book. Identify the type of element by placing a tick in the correct column.

	Name	Metal	Metalloid	Non-metal
a	silicon			
b	carbon			
c	oxygen			
d	iron			

6 Copy the columns below into your exercise book. Match the mixtures with the method of separation by drawing a straight line between them. Each method may be used only once.

	Mixture	Separation method chosen
a	iron and gold	stir with water, one dissolves
b	sulphur and iron	cannot be separated, it is a compound
c	carbon and salt	dissolve one in acid
d	zinc sulphide	magnet

7 What are the simplest numerical ratios of elements in each of the following compounds?

a Pb_3O_4 b $C_{12}H_{22}O_{11}$

c P_4O_{10} d H_2O_2

8 a Use the following experimental data to deduce the ratio of sulphur to the other element in the compounds.

	Mass of element (g)	Mass of sulphur (g)
i	50	15.6
ii	50	23.4

b If compound (i) contains two sulphur atoms for each one of the other element, how many sulphur atoms are there in compound (ii)?

9 Copper oxide (CuO) reacts with hydrogen when heated to leave a brown solid and a colourless liquid.

a Identify the products.

b Write a symbol equation for the reaction.

c When the still hot product is exposed to air, it turns black.
Suggest a reason why.

Learning Checklist

4

☆ I know how many elements there are.	Page 80
☆ I know that atoms can join up.	Page 80
☆ I can describe the differences between solids, liquids and gases.	Page 85
☆ I know that the Periodic Table displays the elements.	Page 85
☆ I know that we use special symbols for the elements.	Page 85

5

☆ I know the difference between atoms and molecules.	Page 80
☆ I know some examples of common chemical symbols.	Page 82
☆ I know some differences between metals and non-metals.	Page 84
☆ I can correctly use the words brittle, malleable and ductile.	Page 84
☆ I know that the Periodic Table shows patterns of elements.	Page 85
☆ I know how compounds and mixtures are different.	Page 91
☆ I can recognise some chemical symbols.	Page 93

6

☆ I know some examples of molecules.	Page 80
☆ I can identify elements as metals or non-metals.	Page 84
☆ I can use the Periodic Table to classify elements.	Page 85
☆ I can distinguish between compounds and mixtures by their properties.	Page 90
☆ I can write chemical symbols from names.	Page 92

7

☆ I can count the atoms in the formula of a molecule.	Page 80
☆ I know the characteristic properties of metals and non-metals.	Page 84
☆ I know the symbols of many common elements.	Page 85
☆ I know that there is a gradation in properties in the Periodic Table.	Page 85
☆ I know in what ways the properties of mixtures differ from those of compounds.	Page 91
☆ I can understand how to use word equations.	Page 93

8

☆ I can convert word equations into symbol equations.	Page 93
☆ I can calculate and explain mass changes in reactions.	Page 95
☆ I can predict whether the mass will increase or decrease in a particular chemical reaction.	Page 95

Topic Quiz

1. Which is simpler, an element or a compound?
2. What do we call the shorthand for elements?
3. What does it mean if something is brittle?
4. Which is more brittle, plasticine or glass?
5. What sorts of materials are displayed in the Periodic Table?
6. What are the **two** major categories of elements?
7. Which is easier to separate, a compound or a mixture?
8. What are the **three** states of matter?
9. What is the smallest possible number of atoms in one molecule?
10. How would you write a sulphur molecule containing eight atoms?
11. How many compounds are there in salty water?
12. How many different elements are there in glucose, $C_6H_{12}O_6$?

True or False?

If a statement is false then rewrite it so it is correct.

1. There are three states of matter.
2. Metals such as gold and copper are elements.
3. The building blocks of elements are molecules.
4. Malleable means bending without breaking.
5. Ductile materials are good for wires.
6. You can separate a mixture of iron and nickel using a magnet.
7. Bromine and mercury are both liquid elements.
8. Most elements are non-metals.
9. Elements always combine in fixed proportions by mass.
10. Water is an element.
11. Chemical symbols only use capital letters.
12. The symbol Au means the element gold.

Literacy Activity

Write a paragraph as if you were a water molecule evaporating from the sea, falling as snow and ending up in a glacier.

ICT Activity

Choose **15** elements from the Periodic Table. Type a table giving their names and symbols and whether they are metals or non-metals.

Magnetism and navigation

FIGURE 1: A 19th Century ship's compass.

The magnetic compass has been used by navigators for thousands of years. By taking bearings of visible objects with a compass, navigators know the position of their ship.

MAGNETISM

- The first people to know about magnetism were probably the Ancient Greeks who lived near the city of Magnesia, which explains where the word magnet came from. It is said that a Cretan shepherd called Magnes first noticed that the iron-tipped end of his crook was pulled down when he walked near to a certain type of rock. The rock was lodestone and it is iron-rich with a natural magnetism.
- It is likely that the early Chinese also knew about lodestones. (Scientists now believe that these lodestones were formed when chunks of iron ore were struck by lightning.) The Chinese probably knew that an iron needle becomes magnetic when it is repeatedly stroked with a lodestone and that the needle, when freely suspended, points north-south.
- As early as the 12th Century simple compasses began to be used and they were in general use by the Middle Ages. A compass is a long slim magnet that can turn or swivel freely. Each of its poles attracts Earth's opposite magnetic pole. This makes a compass line up with Earth's magnetic field and point north-south.
- In about 1600 an Englishman called William Gilbert suggested that the reason compasses always pointed north was because Earth itself acted like a giant magnet.
- Today, a ship anywhere in the world can check its exact position by using a signal from a satellite in orbit. Despite this, all navigators still have a compass on board. Tracy Edwards, who captained the yacht Maiden in the 1989–90 Whitbread Round-the-World Yacht Race, used Navsat (satellite navigation) and found it had so many technical problems that she often used a magnetic compass instead.

BIG IDEAS

By the end of this unit you'll be able to describe magnetic attraction and repulsion, field patterns and electromagnetism. You'll be able to explain how generators are used to provide electricity. You'll be able to compare various energy sources that are used for generation and explain the implications.

What do you know?

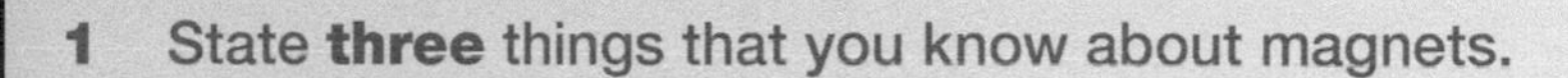

1 State **three** things that you know about magnets. 

2 a What did navigators rely on to find their way before compasses were invented?

  b Would there be any problems with the method in **2 a**?

3 Who were the first people to know about magnetism?

4 Where does the word 'magnet' come from?

5 What do lodestones do?

6 What other metals apart from iron may be attracted to a magnet?

7 What direction does a magnet point in if it can rotate freely?

8 Why do magnetic compasses not point exactly north?

9 Why can ships with iron hulls sometimes cause a compass to become unreliable?

10 What is a magnetic material? Give **one** example.

11 Which parts of a bar magnet are the strongest?

12 Can a magnet attract an object without touching it? Explain your answer.

13 What happens when two bar magnets are placed close together?

Magnetic materials

BIG IDEAS

You are learning to:

- Identify the difference between **magnetic materials** and non-magnetic materials
- Identify magnetic fields and where they are strongest
- Describe some of the uses of permanent magnets

4

Using magnets

Magnets can **attract** (pull) or **repel** (push away) other magnets. They can do this without touching because the magnetic field from each one extends outwards from the magnet.

Permanent magnets keep their magnetism for a long period of time. Most permanent magnets are made from iron although some are made from different materials, but they all contain one or more of the naturally occurring magnetic materials.

There are many different uses of permanent magnets. They are:

- in compasses which are used to navigate.
- in computer hard-drives where data is recorded on to a thin magnetic coating.
- on credit cards, which have a magnetic strip on them to hold information.
- in loudspeakers and microphones, which rely on magnets to work.
- in food manufacture, where magnets are used to keep small metal filings from machinery from getting into food.

FIGURE 1: Some common uses of a magnet.

At home we use magnets to hold things up or to pick up small things.

All of these uses rely on the fact that once a material is magnetised it retains this magnetism for a long period of time.

1. What is a permanent magnet?
2. Give **three** uses of permanent magnets.
3. Explain why it is useful for food manufacturers to use magnets.
4. Why is it useful to have a magnetic strip on a credit card?

... attract ... iron ... magnet ... magnetic field ... magnetic force

Magnetic force

All magnets exert a force on the area around them. The space around the **magnet** which is affected by this **magnetic force** is called the magnetic field. This **magnetic field** cannot be seen with the naked eye. The size of the magnetic field depends on the strength of the magnet.

All magnets have two ends or **poles**. These are called the north seeking (N) pole and the south seeking (S) pole. The magnetic field is strongest at these poles because that is where the strongest magnetic force acts. The further away from these poles, the weaker the magnetic force becomes.

attraction between opposite poles

repulsion between the same poles

FIGURE 2: Like poles repel, unlike poles attract.

If it can turn freely then the north pole of the magnet will turn towards the North Pole of the Earth. The south pole of a magnet will turn towards the South Pole of the Earth. This is how a compass works.

5 Give several factors that might affect how strongly magnets are attracted or repelled to each other.

6 How could you show that a loud speaker has a magnet in it?

How Science Works

It takes much less energy to recycle **steel** than it does to make new steel from **iron**. Because steel is magnetic, magnets are used at recycling plants to separate the steel from other metals. The steel is attracted to the magnets. In 2005 3 billion steel cans were recycled in this way, but several billion were still sent to landfill.

Did You Know...?

- Earth's magnetic north pole is near Bathurst Island in northern Canada, more than 1000 km from the geographic north pole.
- Earth's magnetic south pole is near Wilkes Land in Antarctica, more than 2000 km from the geographic south pole.

Magnetic fields

BIG IDEAS

You are learning to:

- Explain about the shapes, sizes and directions of magnetic fields
- Use magnetic field lines to describe a magnetic field 
- Plot the magnetic field around a bar magnet by using a plotting compass 

Showing the magnetic field 

A **magnetic field** is the area around a magnet in which its **magnetic force** works. The magnetic field of any magnet is stronger at the poles of the magnet. A magnetic object which is put within this field will be affected by the magnet. Magnetic materials will always be **attracted**, but another magnet may be attracted or **repelled**. Magnetism is an example of a non-contact force.

When **iron filings** (very small shavings of iron) are placed near a magnet, they arrange themselves to show the magnetic field. Each small piece of iron has become a very small magnet. They can be used to show how strongly each part of the large magnet attracts them.

This shows that the iron filings are attracted to the magnet. More filings are attracted to the poles of the bar magnet because the field is strongest at these points. In Figure 1, where on the magnets are the magnetic fields strongest?

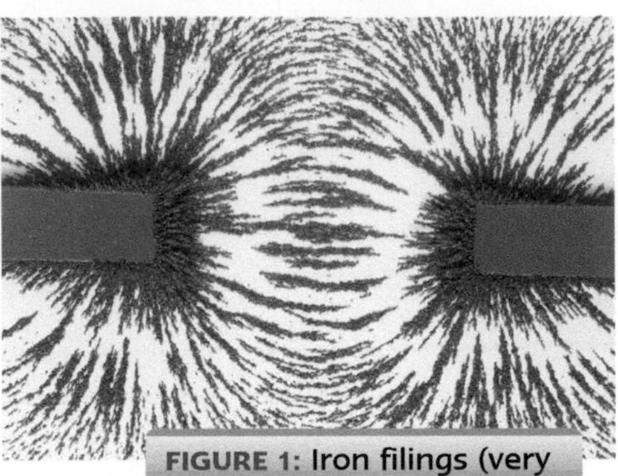

FIGURE 1: Iron filings (very small shavings of iron).

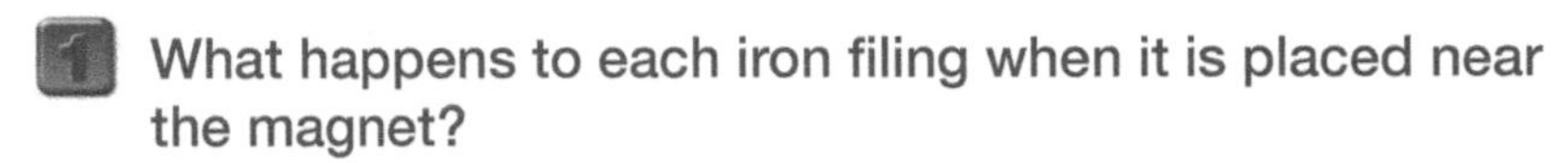

1. What happens to each iron filing when it is placed near the magnet?

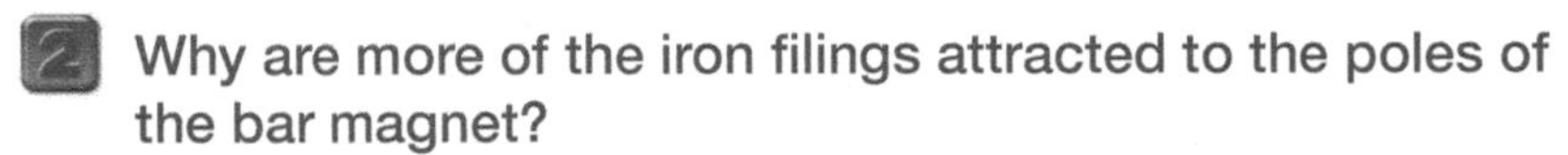

2. Why are more of the iron filings attracted to the poles of the bar magnet?

3. Explain why aluminium filings could not be used to show the magnetic field.

How Science Works

Magnets on a nano scale are set to revolutionise medicine and drug delivery.
By attaching drugs to magnetic particles, a magnetic field can then be used to guide them to exactly where they are needed in the body.

Magnetic field lines

Magnetic fields can also be shown by drawing **magnetic field lines**. When iron filings are placed around a magnet, the filings line up along the field lines. At any point, the direction of the magnetic field is the same as the direction of these field lines. The stronger the field is the less space there is between the field lines. Different kinds of magnets produce different patterns of field lines.

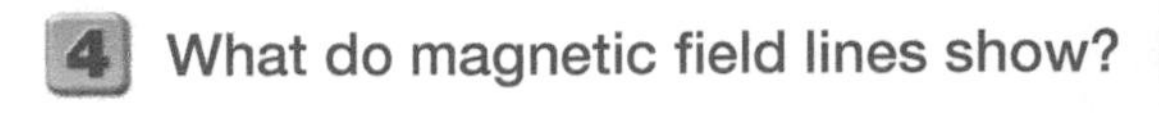

4. What do magnetic field lines show? 

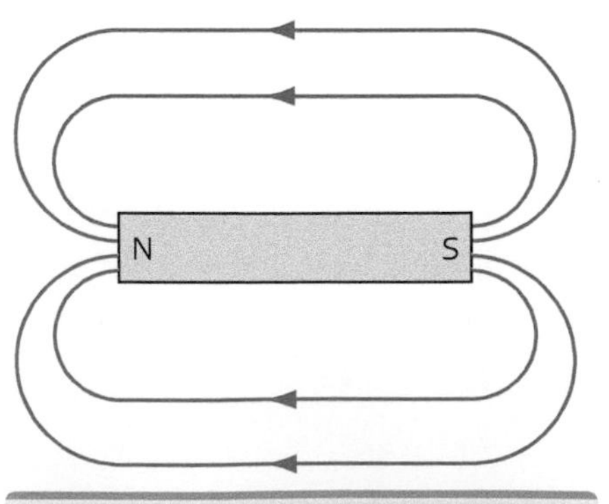

FIGURE 2: Why are the magnetic field lines closer together at the poles of the magnet?

Science in Practice

Investigating a magnetic field

A compass needle is a small magnet. It can be used to investigate the shape and size of a magnetic field that exists around a permanent magnet. A compass needle has a north pole and a south pole. The north pole is shown by the pointed or red end of a compass needle.

The compass needle, which is free to rotate, will line up in the same direction as the field lines. You are going to investigate the magnetic field around a magnet. To do this you will use a compass to plot the direction of the magnetic field at different points around a permanent magnet.

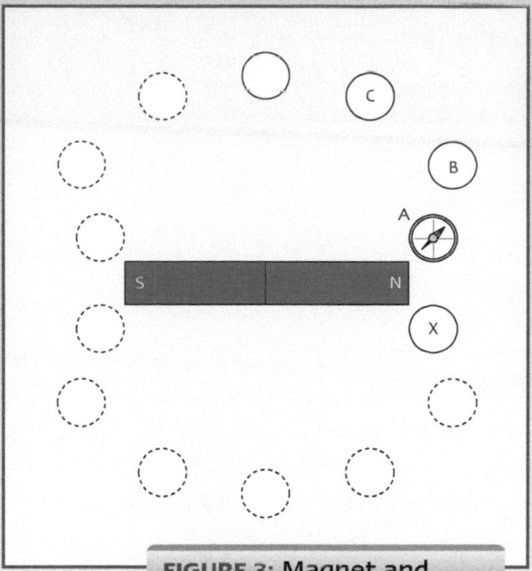

FIGURE 3: Magnet and compass positions.

Method:

1. Place a magnet in the centre of some paper and draw a line around it.
2. Place the plotting compass near to the north (N) pole of the magnet as shown at A in Figure 3 and draw a circle around it. Make one mark at the head of the compass needle and one mark at the tail.
3. Move the compass to position B and make two more marks, one for the head of the needle and one for the tail.
4. Repeat in position C and continue until the compass has been in all the positions on the diagram.
5. Join up all the dots you have marked to make a line. Put an arrow on the line to show which way the compass head was pointing each time.
6. Repeat the process on the other side of the magnet. Start with the compass at position X.

Questions

1. What are the lines you have drawn called?
2. In which direction do these lines always point?
3. What does this tell us about the direction of a magnetic field?
4. Your teacher has given you three metal bars. One is made from iron, one is made from aluminium and one is a magnet. Using only a plotting compass, explain how you could identify each of them.
5. What do you think would happen to the compass needle placed at the midpoint between two magnets with like poles facing each other? Explain your answer fully.

... magnetic field line ... magnetic force ... repel

Earth's magnetic field

BIG IDEAS

You are learning to:
- Explain how a magnetic field goes around the Earth
- Explain what causes this magnetic field and some of its effects

Earth acts like a bar magnet

Compasses point towards the north pole of Earth. This is because the Earth behaves as though it has a large **permanent magnet** at its centre. The invisible force of this huge magnetic field spreads out into space all around Earth. It protects us from **cosmic rays** and the **solar wind** and makes life possible.

1 Why is the Earth's magnetic field so important for life on Earth?

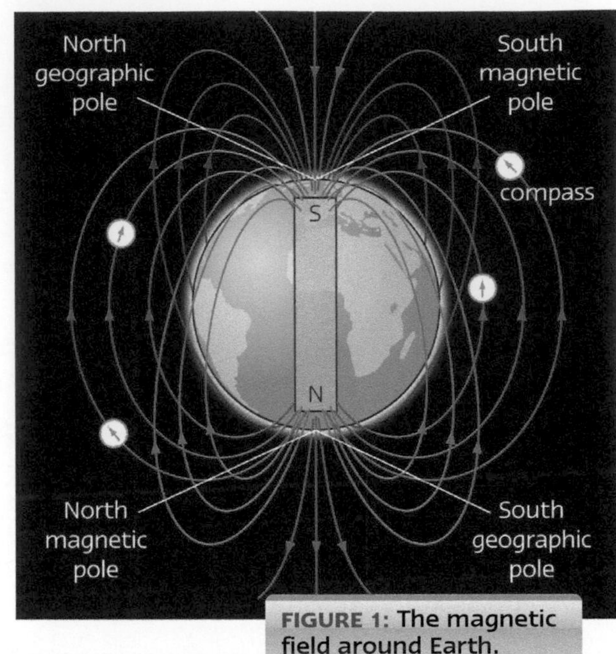

FIGURE 1: The magnetic field around Earth.

What causes the magnetic field around Earth?

Because the north pole of a compass points towards a magnetic south pole, Earth's magnetic field acts as though there is a giant magnetic south pole at the geographic North pole.

Scientists believe that Earth has at its centre a dense fluid **outer core** that surrounds a solid **inner core**, both of which have iron-rich materials in them. The inner core is under such high pressure that it remains solid. The outer core is so hot that it is molten (liquid).

But Earth's iron core cannot be a huge permanent magnet because iron loses its magnetism when it gets too hot.

2 How can we show that the geographic North pole behaves as a magnetic south pole?

3 How do we know that there is not a huge permanent magnet at Earth's core?

The Geodynamo theory

The best explanation for the cause of Earth's magnetic field is the geodynamo theory. This suggests that it is the movement of molten iron in the outer core (caused by the Earth's rotation), combined with convection currents (caused by the heat of the inner core), which create a magnetic field in the molten iron.

... Aurora Borealis ... cosmic ray ... inner core ... magnetosphere

Earth's north and south poles attract charged particles from the Sun. These charged particles collide with gas particles in Earth's atmosphere to create the **Aurora Borealis** (the Northern lights) that are visible within the Arctic circle.

4 What is the geodynamo theory? What evidence would scientists need to obtain to prove or disprove this theory?

5 What causes the Northern lights? Why are they mostly visible at the poles?

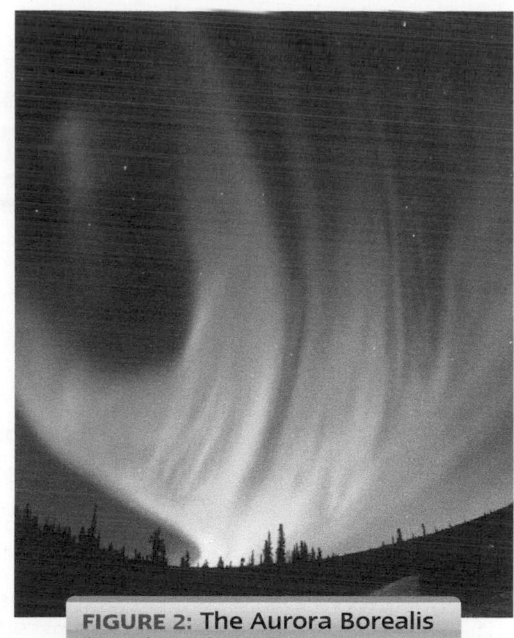

FIGURE 2: The Aurora Borealis (Northern lights).

The Magnetosphere

Earth's magnetic field extends out into space to form the **magnetosphere**. The magnetic field is shaped by the effects of the solar wind. The solar wind is caused by the Sun's activity.

FIGURE 3: The effects of solar wind on the shape of Earth's magnetosphere.

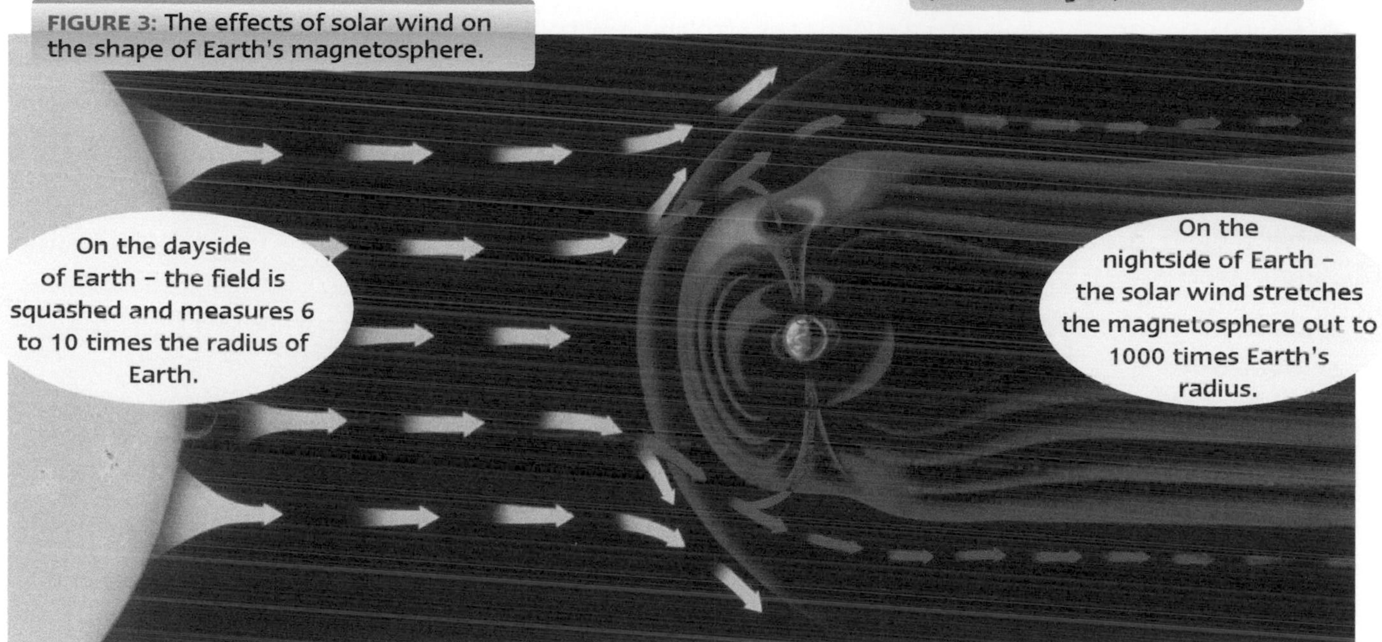

Does Earth's magnetic field change?

Recent evidence from studies of ancient rocks and seafloors suggests that Earth's north and south magnetic poles actually swap around. These reversals have occurred at intervals of about 300 000 years. Analysis suggests that we are overdue another one as the last one occurred about 780 000 years ago. Earth's magnetic field has also weakened by about 10% in the last 200 years.

6 Why is the magnetosphere much larger on the nightside of Earth than on the dayside?

7 What evidence suggests that Earth's magnetic field may swap around in the future? How could this evidence have been obtained?

Watch Out!

Remember that what is called magnetic north when we are using a compass is actually the magnetic south pole of the Earth.

Explaining magnetism

BIG IDEAS

You are learning to:

- Describe how to remove magnetism from magnetic materials
- Describe ways in which materials can be magnetised
- Explain magnetism using the domain theory

Making a permanent magnet

Some materials are so easily magnetised that just holding a **permanent magnet** near to them will turn them into magnets. These materials are called **soft magnetic materials** and soft iron is one of them. Iron objects become weak magnets when they are placed near strong magnets.

Although iron is easier to magnetise, steel stays magnetised for longer.

- A steel rod can be made into a magnet by stroking it with a permanent magnet a number of times in the same direction.
- Placing an iron or steel item in a magnetic field will result in it retaining some of the magnetism on removal.
- Placing an iron or steel item in a coil of wire or **solenoid** which has a direct current passing through it. Direct current flows in one direction only.
- Placing a steel bar in a magnetic field, then heating it to a high temperature and then finally hammering it as it cools.

1. What is a soft magnetic material?
2. Give **three** ways to make a piece of soft iron magnetic.

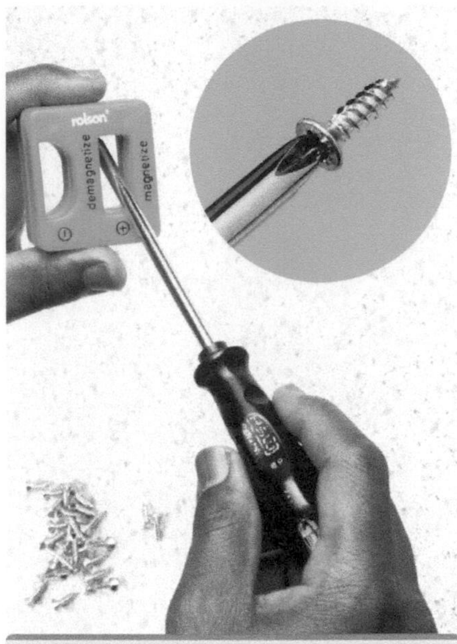

FIGURE 1: How can a steel rod be made into a magnet?

Explaining magnetism 

Little groups of atoms in a magnetic material (iron, nickel and cobalt) behave as tiny magnets. Each of these little groups of atoms is called a **magnetic domain**.

- In an unmagnetised piece of iron, these magnetic domains are arranged randomly and point in lots of different directions. They cancel each other out.
- In a magnetised piece of iron, all these domains point in the same direction. This makes one end of the magnet act as a north pole and the other end act as a south pole. The better the domains are aligned, the stronger the magnet.

How Science Works

The strongest magnet in the world is at Florida State University. It is almost 16 feet tall, has a mass of 15 tons and has taken 13 years to build. It is used for conducting chemical and biomedical research.
The new machine generates a force 420 000 times stronger than the Earth's magnetic field.

... *demagnetised* ... *magnetic domain* ... *non-magnetic materials*

Even when the atoms are oriented randomly, exposure to a nearby magnet may cause them to line up with the magnetic field. Substances that do this easily, such as iron or nickel, can be magnetised.

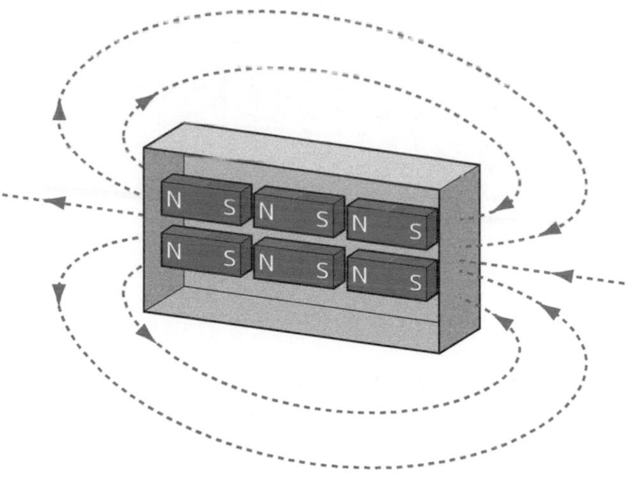

A piece of magnetised iron attracts an unmagnetised nail because it makes more of the domains that are present in the nail line up. When the magnet is removed, most or all of the domains return to a random arrangement within the nail. The nail becomes **demagnetised**.

Non-magnetic materials are those in which the domains remain in a random pattern when they are placed within a magnetic field.

3 What is a magnetic domain? 

4 How is the arrangement of domains different in a magnetised and a non-magnetised piece of iron?

5 Explain why a magnet attracts an iron nail.

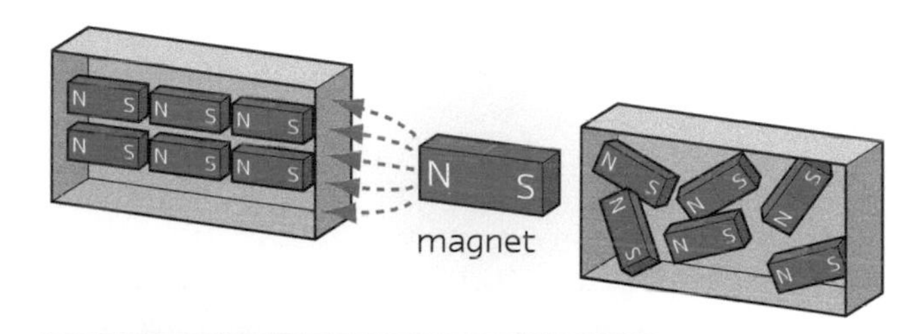

FIGURE 2: What happens to the atomic arrangement in a magnetic material when it is magnetised?

Removing magnetism

In the same way that it is sometimes easy to magnetise a material, it can be equally easy to demagnetise a magnetic material. There are several ways in which this can be done:

- When a magnet is dropped or hammered, the alignment of the domains is reduced and the magnet becomes weaker.
- Heating a magnet.
- Stroking one magnet with another in a random fashion may demagnetise the magnet being stroked. Sometimes the magnetic field is too strong to be removed in this way.

6 Why does dropping a magnet reduce its strength?

7 Explain why a coil of wire with a direct current passing through it will magnetise a piece of iron.

8 Explain why heating a magnet may cause it to become demagnestised.

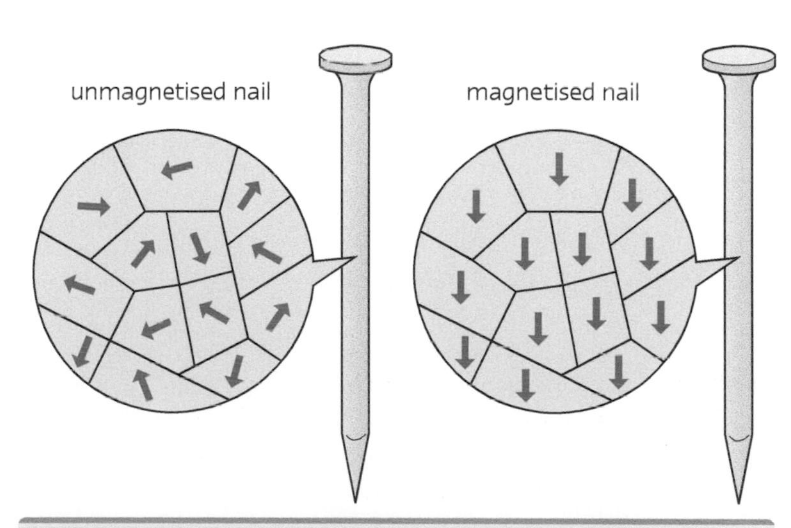

FIGURE 3: What happens to the atomic arrangement in the nails when the magnet is removed?

Electromagnetism

BIG IDEAS

You are learning to:
- Recognise the link between magnetism and electricity
- Describe the magnetic field that surrounds an electromagnet
- Describe how to change the strength of an electromagnet

A simple electromagnet

Passing a **current** through a conducting wire creates a magnetic field around the wire. When the current is switched off the magnetic field goes. A magnet that is made by using electricity is called an **electromagnet**.

1. What happens to a compass needle when it is placed near a wire that has a current flowing through it?
2. What happens to the compass needle when the current in the wire is switched off?
3. What is an electromagnet?

Did You Know...?

In 1820 a Danish scientist called Hans Oersted discovered **electromagnetism.** He placed a compass near to a piece of wire and then passed an electric current through the wire. He noticed that the compass needle moved. When he switched the current off the compass needle went back to its original position.

Making an electromagnet stronger

Electromagnets can be very strong and can be switched on or off. They are only magnetic when the current flows. Switching off the current stops the magnetism.

When a current flows through a single wire the magnetic field around it is very weak. The field can be made stronger by:

- increasing the current passing through the wire
- making the wire into a **coil**
- increasing the number of coils of wire
- placing an iron core in the centre of the coil.

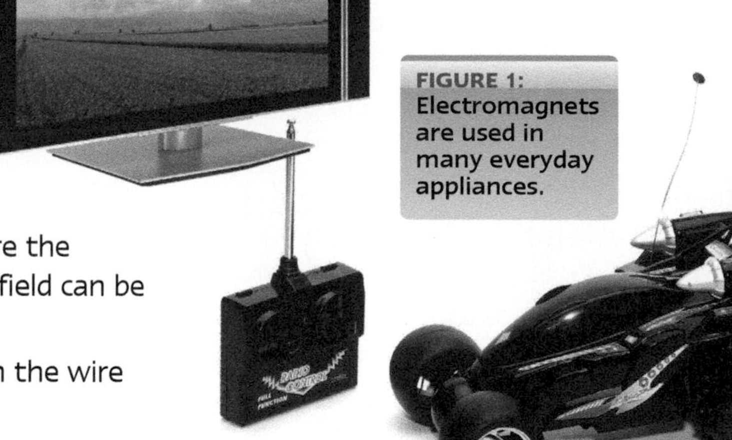

FIGURE 1: Electromagnets are used in many everyday appliances.

Most electromagnets are made by coiling wire around an iron core. When a current passes through the coil, the iron core concentrates the magnetic field and becomes magnetised. Changing the size of the current changes the strength of the electromagnet.

Watch Out!

Remember that an electromagnet will only work when it is part of a complete circuit and the current is flowing through it. If there is no current there is no field produced.

... coil ... current

4 Why are electromagnets often more useful than permanent magnets?

5 Give **three** ways to increase the strength of an electromagnet.

6 Why do most electromagnets have an iron core?

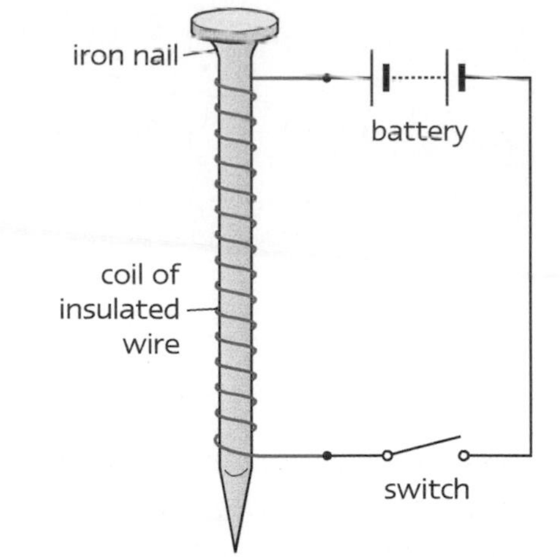

FIGURE 2: A simple electromagnet.

The magnetic field around a coil

Exam Tip!

You are not expected to remember the direction of a magnetic field around a wire or a coil.
But you will need to remember that if the current reverses then the field direction also reverses.

The magnetic field around a coil of wire has a shape which is similar to the shape of the field around a bar magnet. One end of the coil behaves like the north pole of a permanent magnet and the other end of the coil behaves like the south pole.

Reversing the current direction reverses the direction of the magnetic field.

If an alternating current is used then the magnetic field changes direction every time the current changes direction.

7 What factors may affect the strength of the magnetic field generated around a coil?

8 Describe what would happen to the magnetic field if a higher current was flowing through the coil.

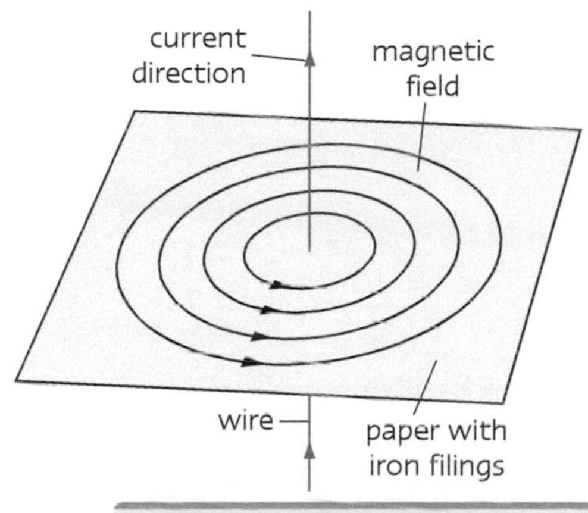

FIGURE 3: Reversing the current direction reverses the direction of the magnetic field.

The magnetic field around a current carrying wire

When plotting compasses are placed around a wire they all point north. When the current is switched on the compass needles move. The magnetic field shape that they show is circular. If the direction of the current is reversed then the direction of the circular magnetic field is also reversed.

Making the wire into a loop increases the magnetic field around the loops. This makes it much stronger and explains why most electromagnets have a coil of wire. By using a magnetic material to make the core, the field is further concentrated.

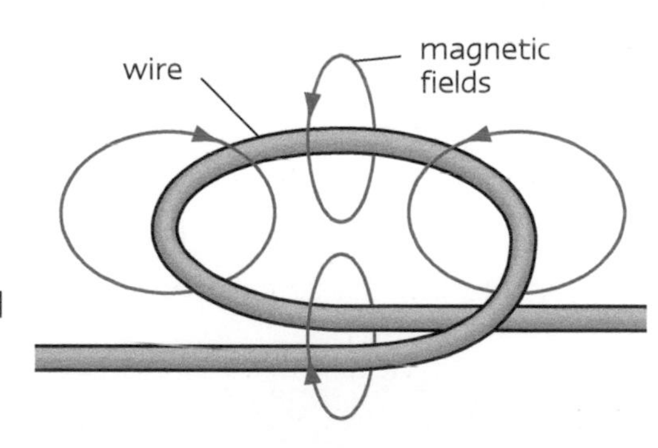

FIGURE 4: The magnetic fields around a loop of wire.

9 Why do the compasses point north before the current is switched on? Which direction will they point when the current is switched off?

10 Explain why making the wire into a loop or coil increases the strength of the magnetised field.

... electromagnet ... electromagnetism

Using electromagnets

BIG IDEAS

You are learning to:

- Describe some of the uses of electromagnets
- Explain how an electric bell, a relay and a circuit breaker work

Electromagnets in industry

Electromagnets have many uses. The main advantage of an electromagnet is the way it can be switched on when a current is switched on. This makes them easy to control. The main disadvantage of an electromagnet is that it needs to be continuously supplied with electricity.

Huge electromagnets are used in scrap yards and in iron and steel foundries.

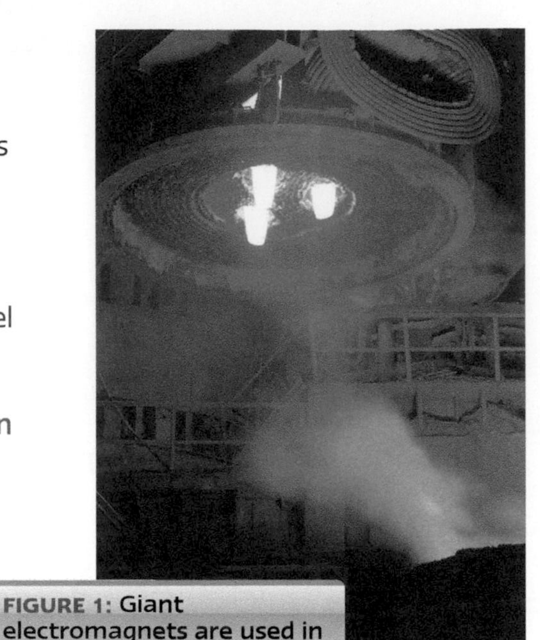

FIGURE 1: Giant electromagnets are used in iron and steel foundries.

1 Why are electromagnets better than permanent magnets in scrapyards?

2 Why is an electromagnet not used in a compass?

The electric bell 

The electric bell works by using an electromagnet to attract a soft iron bar called the **armature**.

1. When the switch is closed a current flows and the soft iron core of the electromagnet becomes magnetised.
2. The armature is attracted to the electromagnet and the hammer strikes the gong.
3. The spring steel strip moves away from the **contact** screw and breaks the circuit.
4. The electromagnet is now demagnetised and the armature moves away from it.
5. The steel strip now touches the contact screw again, completing the circuit and switching on the electromagnet.
6. The cycle then repeats.

3 Why is the armature connected to a springy steel strip?

4 What must be done to stop the bell ringing? Explain your answer.

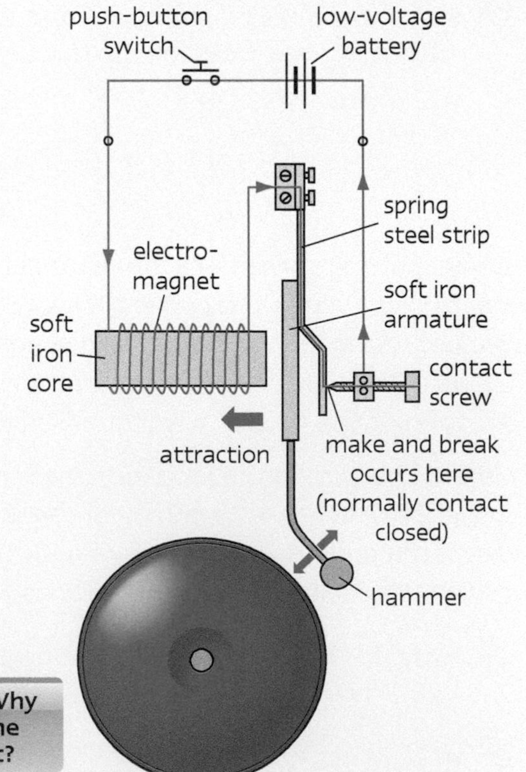

FIGURE 2: The electric bell. Why is there a soft iron core in the centre of the electromagnet?

... armature ... circuit breaker ... contact ... current

The relay

Electromagnetic **switches**, or **relays**, can be used to control different devices. Relays are often used as a kind of safety switch to allow a circuit with a small current flowing through it to switch on a circuit that will have a larger current flowing through it.

Circuit 1 contains a simple electromagnet and only needs a small current. Circuit 2 contains a motor which needs a large operating current.

When the switch is closed a current flows in circuit 1 and this attracts the rocker arm to the electromagnet. The rocker arm pivots and pushes the contacts together. This completes circuit 2 which is switched on. A large current can now flow in circuit 2 to power the motor.

When the switch is opened there is a gap in the circuit, the electromagnet releases the rocker arm and circuit 2 is switched off as the spring moves the contact apart.

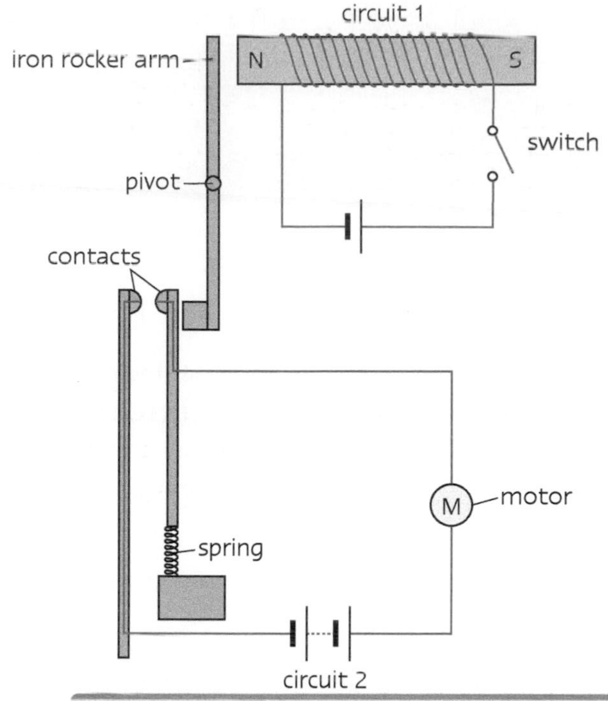

FIGURE 3: Relay circuits like this are used in car starter motors.

5 What are the advantages of using a relay circuit?

6 What is the difference between the two circuits which are connected by the relay?

The circuit breaker

A **circuit breaker** is designed to act as a safety device to prevent an appliance drawing too much **current** from the mains.

If the device is operating normally the low current means the electromagnet is not strong enough to separate the two contacts. But if the device malfunctions and too much current flows through the live wire then the electromagnet increases in strength. The magnetic field is now strong enough to attract the nearest iron contact to it. This then breaks the circuit and the current stops flowing.

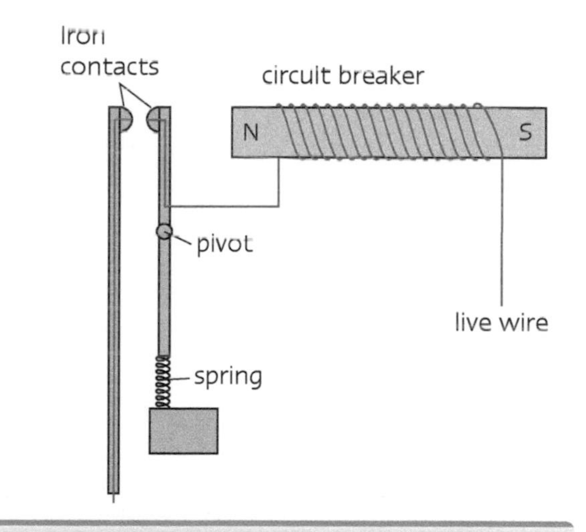

FIGURE 4: A circuit breaker. Why are the contacts made from iron?

The spring then prevents the contacts from reconnecting. When the fault is fixed, the contacts are reset by pressing a switch on the outside of the circuit breaker.

7 How does a circuit breaker prevent too much current flowing into a device?

8 What advantages do they have over a normal on/off type of switch?

9 Where are circuit breakers commonly used at home?

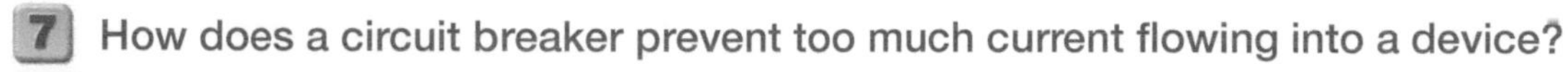

Motors and generators

BIG IDEAS

You are learning to:
- Understand about Michael Faraday and his discovery of the motor effect HSW
- Understand how the motor effect works and some of the uses of electric motors
- Understand how magnetism can be used to generate electricity

Uses of electric motors 

In 1831 the English physicist Michael Faraday found that if a wire carrying a current was placed in a magnetic field then the wire would move. The wire moves because the magnetic field around the wire repels the magnetic field around the wire.

This important discovery is called the **motor effect** and it is used in all **electric motors**. Every electrical device that transfers **electrical energy** into movement or **kinetic energy** uses an electric motor.

1. What did Michael Faraday discover in 1831?
2. Why does a current carrying wire move when it is placed near to the poles of a magnet?

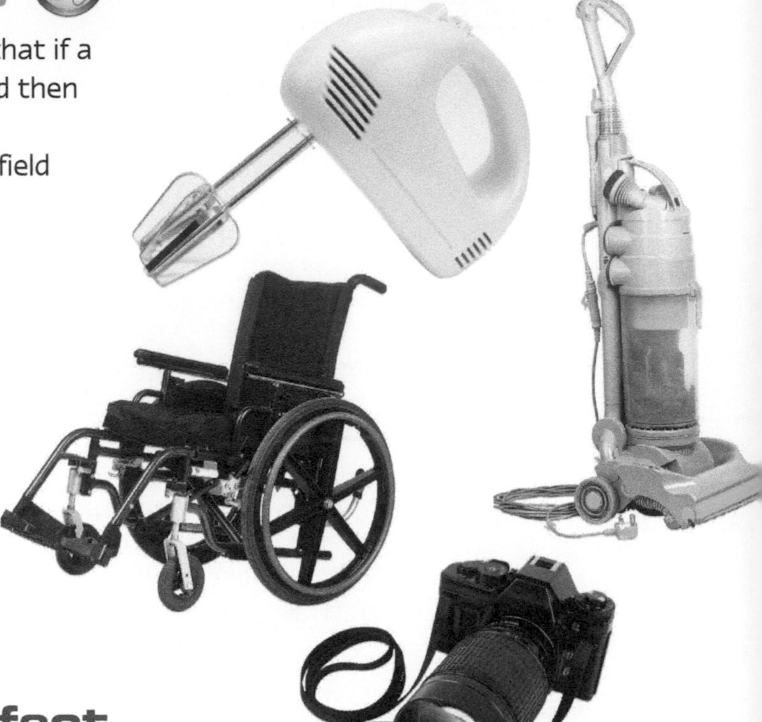

FIGURE 1: Electric motors are commonly found in food mixers, vacuum cleaners, automatic cameras, washing machines, sewing machines and motorised wheelchairs. What energy transfer occurs in all these devices?

Explaining the motor effect

When two magnets are placed close together, their two magnetic fields will interact and they either **attract** or **repel** each other.

So if a wire is placed in the field of a permanent magnet and an electric current is passed through it, then there are now two magnetic fields, one from the permanent magnet and one from the wire. These two magnetic fields will either attract or repel each other and the wire will move.

If there is no current then there is no movement.

If the direction of the current, or the magnetic field, is reversed then the direction of movement also reverses.

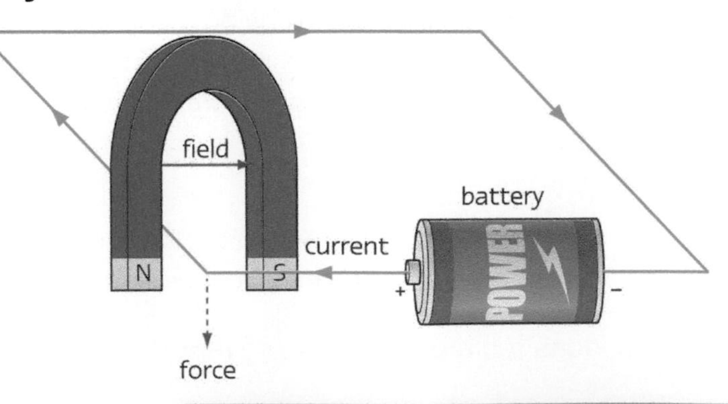

FIGURE 2: The motor effect. What happens if the current in the wire is switched off?

The amount of movement of the wire can be increased by:

- increasing the size of the current passing through the wire
- increasing the size of the magnetic field
- making the straight wire into a coil
- increasing the number of turns on the conducting wire.

Commercial electric motors incorporate all these improvements. They are usually built into a circuit containing a variable resistor to control the size of the current and the speed of the motor. Motors use the non-contact force of magnetism and work by changing magnetic energy into kinetic energy.

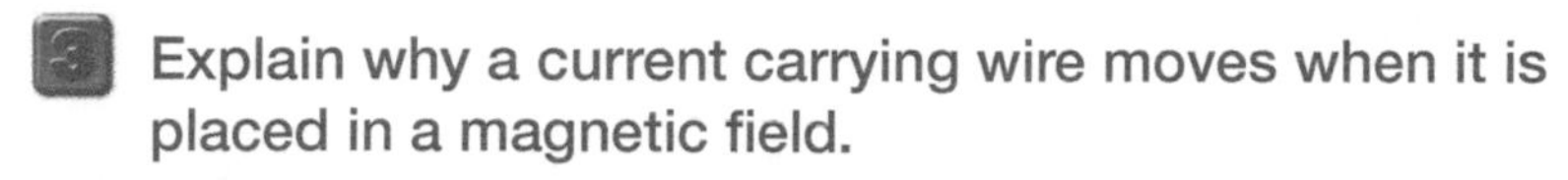

3 Explain why a current carrying wire moves when it is placed in a magnetic field.

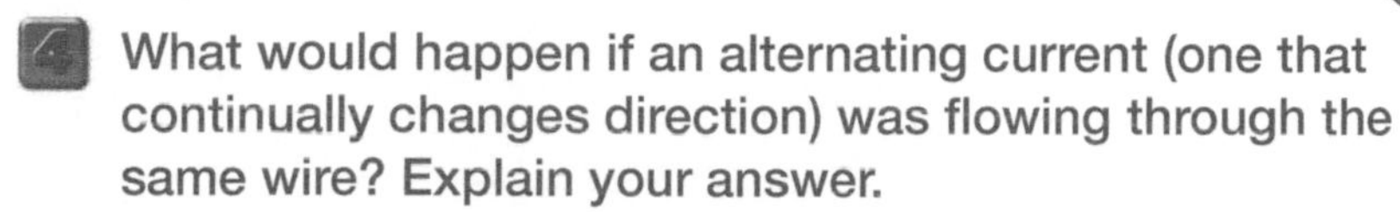

4 What would happen if an alternating current (one that continually changes direction) was flowing through the same wire? Explain your answer.

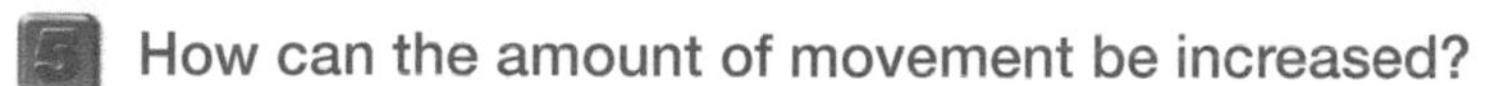

5 How can the amount of movement be increased?

How Science Works

Scientists recently unveiled the tiniest electric motor ever built. The motor works by shuffling atoms between two molten metal droplets in a carbon nanotube.

HSW

Watch Out!

Remember:

- electric motors use an electric current and a magnetic field to produce movement
- generators use a magnetic field and movement to generate an electric current.

The generator effect

Generators use the non-contact force of magnetism to convert kinetic energy into electrical energy. The **generator effect** is rather like the motor effect in reverse. When a magnet is pushed into a coil made from conducting wire, which is part of a complete circuit, a current flows in the wire. (This is called an **induced** current.) There is only a current when the magnet is moving. The current can be increased when:

- using a stronger magnet
- the coil has more turns of wire in it
- the magnet is moved faster.

The current changes direction when the magnet is pulled out or the other pole of the magnet is pushed in.

To keep the current flowing the magnet must be kept moving. One common way to do this is to spin the magnet. Machines that do this are called **dynamos**. Some bicycle lights use dynamos.

6 What is the difference between the motor effect and generator effect?

7 How can the size of the induced current be changed?

8 How can a dynamo be used to power bicycle lights? What are its disadvantages?

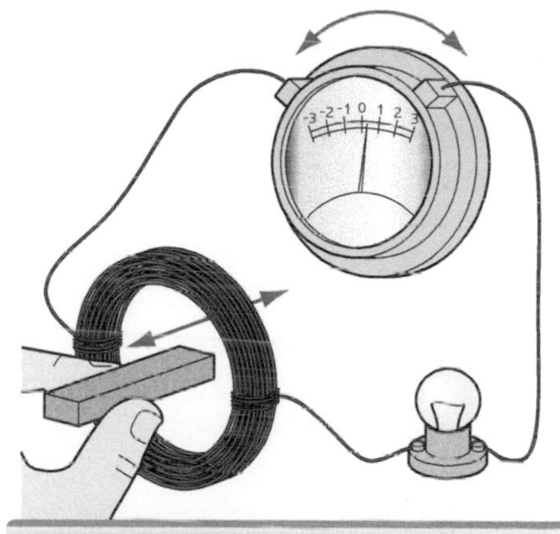

FIGURE 3: The generator effect. What happens to the current if the magnet is kept still?

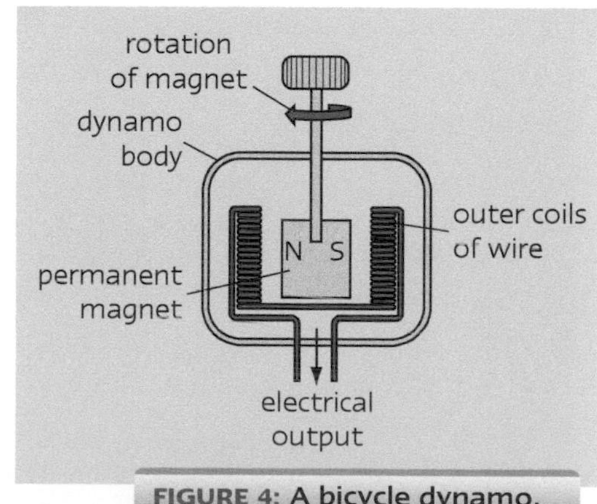

FIGURE 4: A bicycle dynamo.

Maglev Trains

Plans to spend an estimated £1.8 billion on building a 311 mph maglev (*magnetic levitation*) train line connecting Glasgow and Edinburgh have divided local and national opinion. The levitated trains could cut the journey time between Scotland's two major cities to just 15 minutes.

How do Maglev trains work?

Maglev trains differ from normal trains in not having an engine. They rely on an electromagnetic force that suspends (levitates) the train above a guidetrack.

The magnetic field created around a current carrying wire is the simple idea behind a maglev train rail system. There are three components to this system:

- a large electrical power source
- metal coils lining a guideway (track)
- large guidance magnets attached to the underside of the train.

When a current flows in the coils that line the guideway, a magnetic field is induced around them. This field repels the large magnets on the train's undercarriage causing the train to levitate between 1 and 10 cm above the guideway.

By removing friction (there is no contact with the rails as in conventional trains) and using aerodynamic designs, the trains can reach speeds of more than 310 mph (500 kph).

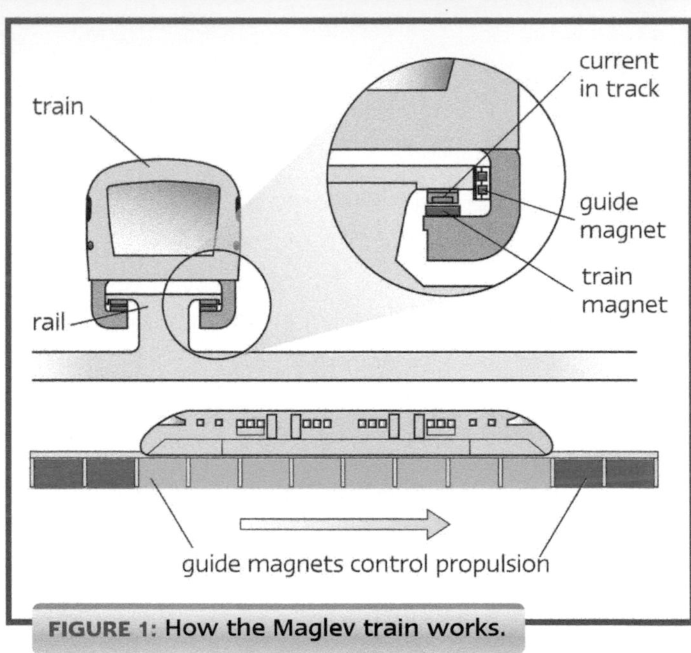

FIGURE 1: How the Maglev train works.

Are there environmental benefits?

Maglev trains have many environmental advantages:

- there is no engine and no wheel contact so there is insignificant noise pollution or vibration
- there is no use of fuel so there is no air pollution (at least not from the vehicle).

Maglev trains have the ability to climb a 10^0 gradient which has led NASA to begin to develop maglev technology with the aim of using it to launch spacecraft.

Assess Yourself

1. What is the main difference between a maglev train and a normal train?
2. What basic system do all maglev trains use?
3. What are the **three** main components of the system?
4. Briefly explain why a maglev train stays suspended above the guideway.
5. What could be done to stop the train (other than using brakes)?
6. Why are maglev trains able to travel much faster than normal trains?
7. Name **two** environmental advantages of maglev trains.
8. Why is maglev technology currently being developed for launching spacecraft?
9. It is likely that most maglev rail systems will be initially developed in countries such as China and India rather than European countries. Suggest why this is the case.

Geography Activity

What would be the main social and economic benefits to the economy of Scotland and in particular to Glasgow and Edinburgh if and when such a rapid transport system becomes available? What issues will need to be considered during the planning and construction phases of the project?

ICT Activity

You are going to research some information about the development of maglev technology over the last 50 years. Try to find out who first developed the idea and if there have been other railways which have used the technology. You will need to find out when the key ideas were developed and provide some facts about them.

Level Booster

8 You can analyse some of the potential problems associated with maglev technology and consider the social, environmental and economic issues associated with the development and introduction of new technologies.

7 Your answers show an advanced understanding of magnetism including maglev technology and an advanced grasp of scientific terminology. You can demonstrate an advanced appreciation of how developing transport mechanisms such as maglev will resolve some current transport problems and recognise how scientific progress can help to solve social and economic problems.

6 Your answers show a good understanding of magnetism, a good grasp of scientific terminology and a good appreciation of how the use of this technology will impact on people's lives.

5 Your answers show a good understanding of magnetism, a good grasp of some scientific terminology and some appreciation of how scientific problems are resolved.

4 You can describe some of the uses of magnetism. Your answers show a basic understanding of magnetism and a basic grasp of some scientific terminology.

Power stations

BIG IDEAS

You are learning to:

- Identify what fuels power stations use
- Understand how power stations work and the impact they have on the environment

Supplying electricity

FIGURE 1: A power station that burns coal. Do you see the large towers? Try to find out what these towers are called and what the white clouds coming out of them are.

Most of the very large amounts of energy that we use in homes, factories and schools is **generated** in **power stations** which burn fossil fuels. There are many different designs of power station but they all work in the same way by burning fossil fuels and converting this energy into **electricity**. The electricity is then taken to where it is needed.

1. What does a power station do?
2. What are the main fuels burnt in power stations?

Inside a power station

There are seven main steps that take place in a power station.

3. Draw a flowchart to explain how a power station converts coal into electricity.
4. Why is it important to **recycle** the water that is used?

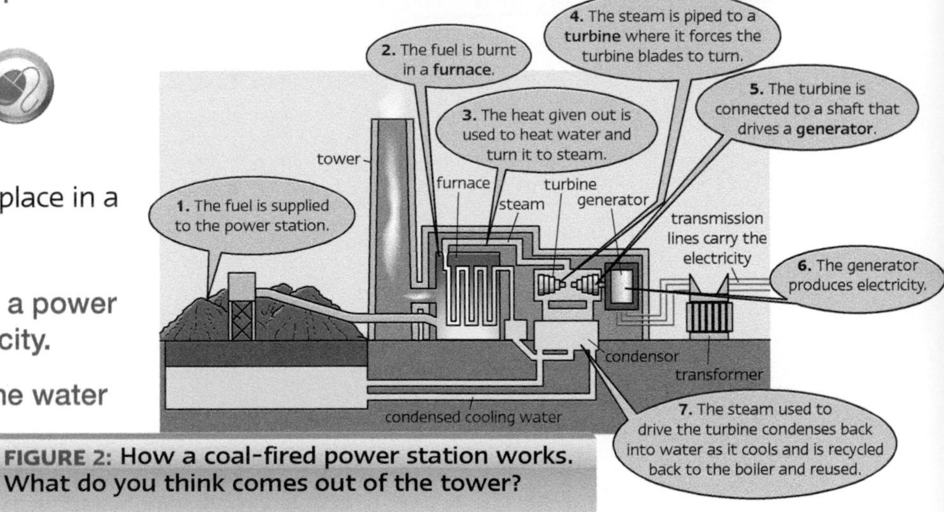

FIGURE 2: How a coal-fired power station works. What do you think comes out of the tower?

Efficiency of power stations

A power station is not very **efficient** at converting the stored energy in fuel into electrical energy. When the power station burns fossil fuels it produces large quantities of water vapour and carbon dioxide. These hot gases leave the power station through its towers. Carbon dioxide cannot be seen but water vapour can be as it is cooled by the surrounding air and condenses. At least 60% of the energy stored in the fuel is lost as heat through the towers. Power stations also lose a lot of energy as heat in the generating process.

5. Where does a power station lose heat?
6. Suggest why a power station loses a lot of energy during its generating process.

FIGURE 3: A turbine in a modern power station. What role do turbines have in the production of electricity?

... alternating current ... efficient ... electricity ... furnace ... generate

Distribution and the National Grid

Most modern power stations are more efficient than older power stations. The electricity that they produce is distributed in cables that may be buried underground or suspended between pylons. All power stations produce **alternating current** (A.C.) electricity at the same frequency. This means that all the power stations can be connected together in a countrywide distribution system called the **National Grid**.

7 Suggest features that a new power station has to make it more efficient than an older one.

8 What is the National Grid?

9 What is the advantage of all power stations producing electricity at the same frequency?

Watch Out!

Electricity is the *product* of a power station. There are a lot of different energy transfers that happen along the way. Remember electricity is produced by generators (*not* turbines) in a power station.

FIGURE 4: Pylons carry electricity wires overhead for safety.

Problems with burning fossil fuels

Burning large amounts of fossil fuels to generate electricity has the following disadvantages:

- Carbon dioxide is a greenhouse gas which contributes to global warming.
- Sulphur dioxide, which is also produced, especially when low-grade fuels are burnt, causes acid rain.
- Transporting fuels to power stations can cause pollution.
- Large quantities of water are used in the cooling process.
- If coal is burnt, large quantities of ash are produced.

To reduce the effect of some of these drawbacks, power stations are often situated close to the coast or near to large rivers. Coal-burning power stations are usually very close to coalfields to reduce the distance that coal has to be transported.

10 What are some of the issues that need to be considered when deciding where to build power stations? Explain your answer.

11 Suggest how the amount of polluting gases entering the atmosphere from power stations can be reduced.

12 What sort of issues may arise from trying to reduce pollution?

How Science Works

Scientists have discovered that the waste ash (called 'fly ash') produced by coal-fired power stations can be used to make concrete. However, it is estimated that there are still 53 000 000 tonnes of unused fly ash in the UK.

HSW

... generator ... National Grid ... power station ... recycle ... turbine

Burning problems

BIG IDEAS

You are learning to:

- Understand the environmental effects of using fossil fuels to generate electricity (HSW)
- Evaluate the risks of our continuing to rely on fossil fuels as an energy resource
- Identify some examples of pollution

Burning fossil fuels

We continue to burn very large amounts of fossil fuels to produce the energy that we use in the world today. There are two main problems with this:

- the supply of these fuels will run out
- it causes **pollution** of the air and water.

We need to make sure we always have clean air and water to use.

1 Why do we need to be careful how much of the fossil fuels we use?

FIGURE 1: Pollution of the air is worse from power stations like this one that burn poor quality coal. (Not all this cloud is polluting gases; some of it is water vapour.)

Atmospheric pollution

Burning fossil fuels releases huge amounts of pollutants into the **atmosphere**. The table shows the main atmospheric pollutants and their effects.

In certain parts of the world this pollution of the air is much worse. Cities where there are large numbers of cars and industry can be especially bad.

Oil refineries also produce large amounts of atmospheric pollution during the petrochemical refining process.

Name of pollutant	Problems it causes
carbon dioxide	• **greenhouse gas**, a cause of **global warming**
sulphur dioxide	• greenhouse gas, a cause of **acid rain**
soot and particulates (tiny particles)	• breathing problems • deposited on buildings turning them black • block air filters
other chemicals	• breathing problems • allergies • **smog** (a layer in the atmosphere where pollutants have become very concentrated)

2 List the main atmospheric pollutants and their effects.

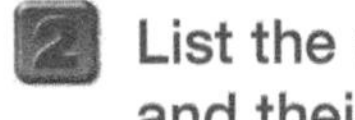

3 What can people living in affected cities do to reduce the risk to their health?

4 What is global warming? Do you think it is a problem?

... acid rain ... atmosphere ... crude oil ... environmental cost

Water pollution

One of the biggest risks with using oil as a fuel is the danger of pollution during its transportation. Because oil is used in such large amounts and often has to be transported over large distances, things can go wrong:

- Pipelines can leak or be destroyed by terrorists.
- Tankers can sink or run aground releasing large amounts of highly polluting **crude oil** into the sea.

Power stations use large amounts of water when they generate electricity. Sometimes chemicals get into the water and pollute it.

5 Why should power stations be careful to ensure that pollutants are not released into the environment?

6 What are some of the dangers of transporting oil?

Did You Know...?

In Tokyo the pollution from car engines is so bad that the traffic police are only allowed to be on duty for 20 minutes. After this time they need to breathe oxygen from a cylinder to help their bodies recover.

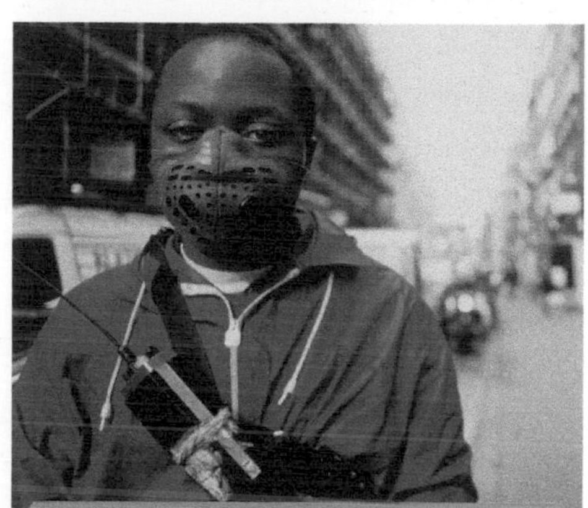

FIGURE 2: A courier working in a polluted city wears a face mask to filter the air he breathes.

During the first Gulf war in 1991, oil was deliberately leaked into the sea and wells were set alight. This caused extensive water and air pollution.

7 Why does oil have to be transported over large distances?

8 List the main effects of oil on wildlife.

9 What other sorts of pollution can oil cause?

Case study – the *Exxon Valdez* disaster

HSW

It is very difficult to clean up after oil spills. A number of different methods are used. Most of these rely on the use of detergents to disperse the oil. Affected wildlife can be cleaned but this is very time consuming and only works in a small number of cases.

One of the worst ever cases of marine pollution was when the supertanker *Exxon Valdez* ran aground in Alaska in 1989. It released 10.8 million gallons of oil into the sea and polluted over 1200 miles of coastline. This had serious effects on local wildlife and started a debate about the **environmental cost** of our continued need for oil.

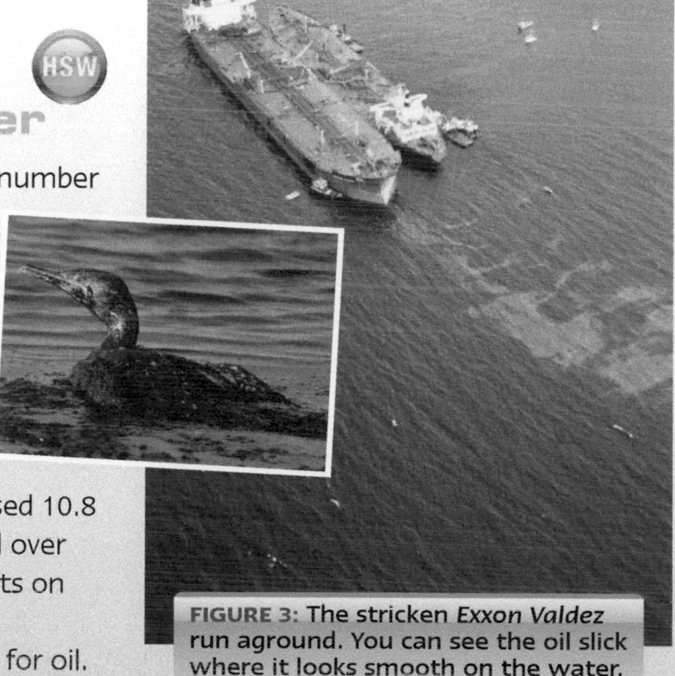

FIGURE 3: The stricken *Exxon Valdez* run aground. You can see the oil slick where it looks smooth on the water.

10 Explain what is meant by the term 'the environmental cost'.

Renewable energy resources

BIG IDEAS

You are learning to:

- Describe some types of renewable energy
- Understand the advantages and disadvantages of these types
- Understand how renewable energy resources can be used to produce electricity
- Explain the implications of using renewable energy resources

Why do we need other fuels?

To make our fossil fuels last longer and to reduce the amount of pollution, we need to find other ways to produce the energy that we need. The best method is to develop ways to produce energy from **renewable resources**. Renewable energy resources are energy resources that will not run out.

One of the best ways is **hydroelectric power (HEP)**. It uses the energy from falling water to turn **generators** that make electricity.

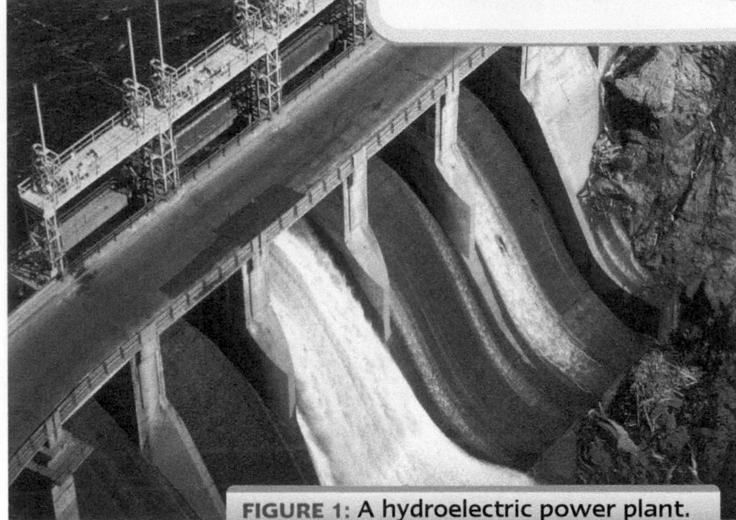

FIGURE 1: A hydroelectric power plant.

1. What does a 'renewable energy resource' mean?
2. What problems might there be in using falling water as an energy resource?

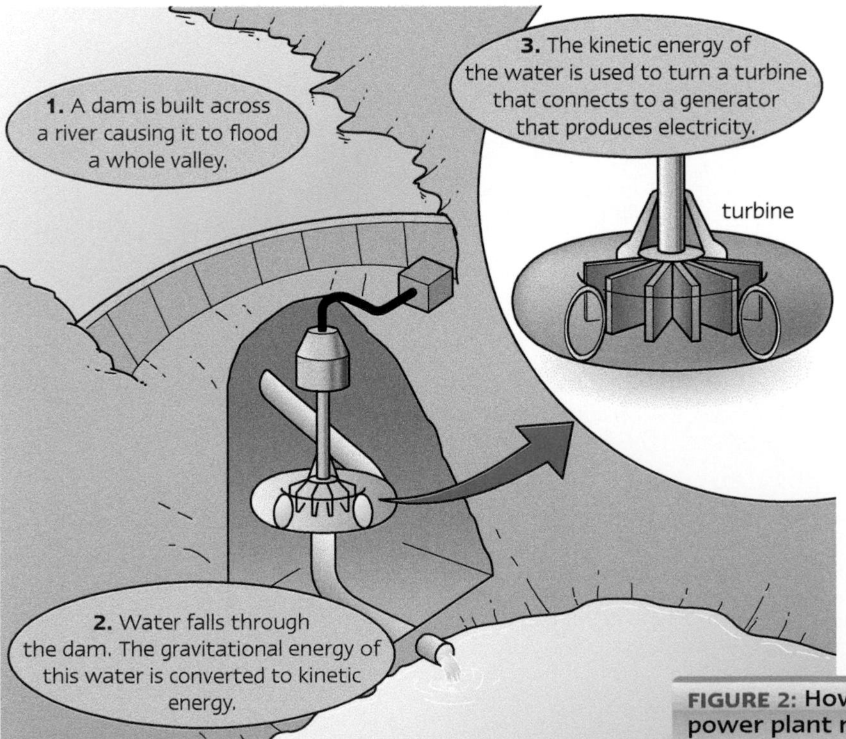

FIGURE 2: How a hydroelectric power plant makes electricity.

Did You Know...?

The largest hydroelectric power plant in the world will be the 'Three Gorges' project across the Yangtze river in China. The dam wall is 185 m (607 ft) high and 2309 m (1.4 miles) long.
Over 1.4 million people will have to be relocated as their homes will be flooded when the dam is made.

... bio-fuel ... generator ... geothermal ... hydroelectric power (HEP)

What are the renewable energy resources?

Renewable energy resources that are currently widely used include the following:

- **Water** energy – stored energy in falling water is used to turn a turbine.
- **Wave** energy – kinetic energy in ocean waves is used to turn a turbine.
- **Wind** energy – kinetic energy in wind is used to turn a turbine. People have used this for centuries.
- **Tidal** energy – as tides rise and fall the seawater passes through a turbine which generates electricity.
- **Solar** energy – panels absorb heat from the Sun and use it to heat water; cells use it to produce electricity.
- **Geothermal** power – heat in rocks inside the Earth is used to turn water into steam to power a turbine.
- **Bio-fuel** – chemical energy in plants is released as the plant materials are burnt.

The energy in all these renewable energy sources, with the exception of bio-fuels, can be used directly (via a turbine or other converting system) – no burning is needed. This reduces pollution.

Exam Tip!

Remember that renewable energy resources are less efficient at generating electricity than traditional coal-fired power stations. They also have a significant environmental impact. For example, wind turbines spoil the view and some scientists say that many birds are killed flying into the turbine blades.

FIGURE 3: A coastal wave power station called 'Limpet 500'. It generates 500 kW of electricity, enough to power 300 homes.

3 Give **four** examples of renewable energy resources that can be used to generate electricity. For each one describe how it works.

4 What do all of your named methods have in common?

How will we meet our energy needs?

Most of these energy resources can only contribute a small amount of our total energy needs. This is because they are far less concentrated energy sources than the energy in fossil fuels (they have low energy densities). Many different types of renewable energy resources will have to be used at the same time to even begin to produce enough electricity for our future needs. Other problems are:

- The sites are inaccessible which makes maintenance difficult.
- It is often difficult to connect remote sites to the National Grid.
- Many renewable energy sites only generate electricity when the conditions are right.
- The machines that use renewable resources to generate electricity are often manufactured using energy from fossil fuels.

5 Why will lots of different renewable energy resources need to be used together in the future?

6 What can prevent renewable resources from generating electricity all the time?

7 What do we need to consider when deciding if these resources represent a viable future?

What about nuclear power?

BIG IDEAS

You are learning to:

- Consider if nuclear power is an effective way of generating electricity in the future
- Describe some of the advantages and disadvantages of nuclear power 
- Explain the implications of using nuclear power 

Why nuclear power?

Even all the renewable energy resources working together cannot meet our future energy needs. Each one of them has advantages and disadvantages. To meet our future energy needs we must consider the use of **nuclear** power.

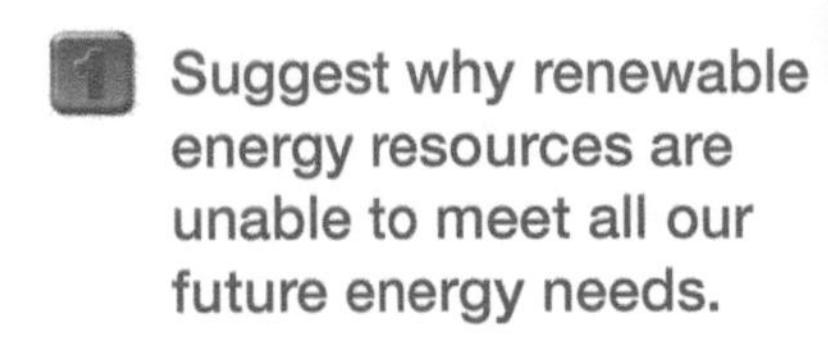

1 Suggest why renewable energy resources are unable to meet all our future energy needs.

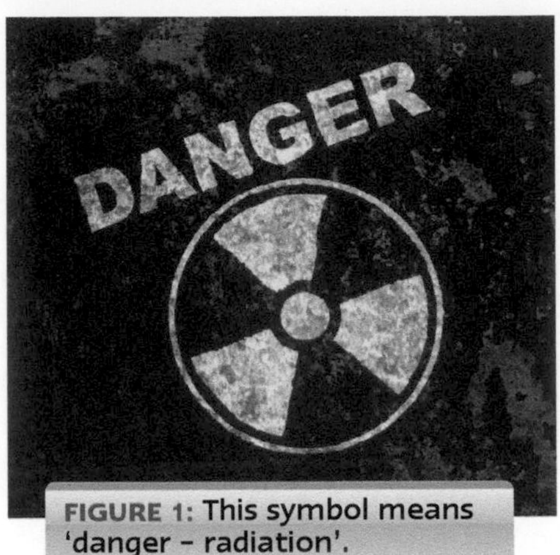

FIGURE 1: This symbol means 'danger – radiation'.

Did You Know...?

One way to dispose of nuclear waste is by burying it deep underground. The waste is encased in tough materials and the surrounding rock acts as a barrier to prevent radiation leaking into the environment. Why do you think these burial sites are not built in earthquake zones?

What is nuclear power?

- Nuclear power is a new and developing technology. It can also be a very dangerous one. Nuclear power was first used in the 1950s but has become more common recently.
- Nuclear power stations use the heat from nuclear reactions to heat water and produce steam used to drive turbines which power generators that produce electricity.
- Nuclear power is used to generate about 5% of the world's electricity supply. This is nearly double the energy produced by all the other renewable energy resources combined.

2 Why do you think nuclear power became more common towards the end of the last century?

FIGURE 2: A modern nuclear power station. Do you think Britain should build more nuclear power stations?

Advantages and disadvantages of nuclear power

Advantages of nuclear power	Disadvantages of nuclear power
Only a small amount of nuclear fuel such as uranium is needed to produce a large amount of energy.	It is a **non-renewable** energy resource because the mineral uranium which is used as the fuel has to be mined from deposits in the Earth.
It lasts a long time.	**Nuclear radiation** can cause cancer and death.
It produces electricity very efficiently.	The technology required is expensive and closely guarded by those few countries which possess it.
It does not contribute to global warming or release gases that pollute the atmosphere.	When the power station reaches the end of its working life it has to be **de-commissioned**. This is a very expensive and dangerous process.

3 Name:

a **two** advantages of using nuclear power to generate electricity

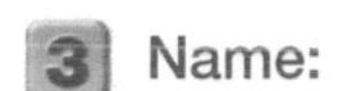

b **three** disadvantages of using nuclear power to generate electricity.

4 Suggest why de-commissioning a nuclear power station is such an expensive and dangerous process.

FIGURE 3: An explosion from an 11-megatonne nuclear bomb. The explosion causes a typical 'mushroom cloud' in the sky.

Potential dangers of nuclear energy

- The same material that is used as a fuel in nuclear reactors is also used to make a **nuclear bomb**. There is always a danger that the uranium will fall into the wrong hands with potentially devastating consequences.
- Nuclear power produces lethally toxic waste which remains active for thousands of years. Currently there is no real solution to the problem of how to dispose of the waste safely.
- It is difficult to find good places to build nuclear power stations.

5 What are the challenges that need to be addressed in using nuclear power?

6 Suggest why it is difficult to find suitable sites to build nuclear power stations.

Practice questions

1 Complete the following sentences in your exercise book. Use words from the list.

attract **attracted** **magnetic** **metals** **repel**

Some .... such as Iron, Nickel, Cobalt and Steel are .... When these materials are placed close to a magnet they are .... to the magnet. When two magnets are placed close together they will either .... or .... each other.

2 Complete the following sentences in your exercise book. Use words from the list.

away **compass** **field** **magnetic**
north **paper** **shape** **south** **weaker**

The space around a magnet which has the .... force in it is called the magnetic .... The magnetic force surrounds the magnet in all directions and becomes .... as you move further from the magnet. The .... of the magnetic field can only be found by covering the magnet with .... and then sprinkling iron filings over the paper. The shape of the field can also be found by using a small plotting .... The direction of the magnetic field is always .... from the .... pole and towards the .... pole.

3 **a** Make a rough copy of the coal-fired power station and place the following labels on it.

generator **boiler (furnace)**
turbine **pylon**

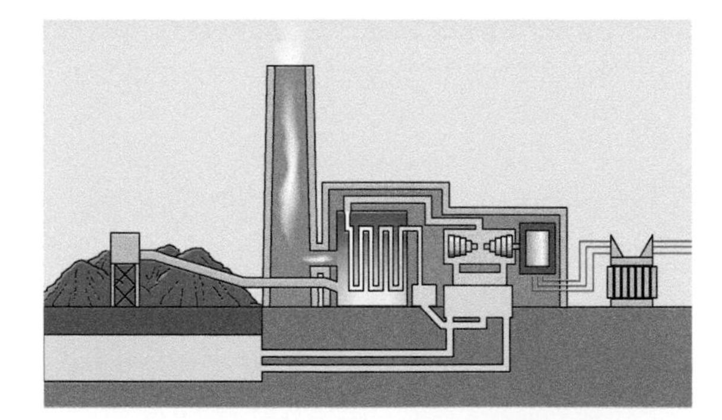

b Put the following events in the order in which they occur.

The steam drives the turbine.
Heat energy produces steam.
The generator produces electricity.
The coal is delivered to the power station.

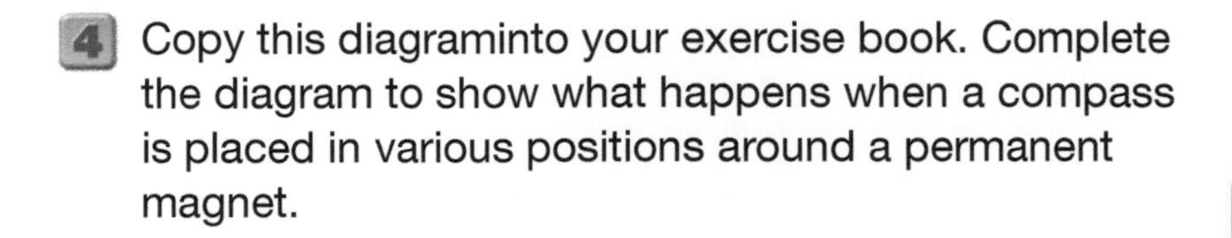

4 Copy this diagraminto your exercise book. Complete the diagram to show what happens when a compass is placed in various positions around a permanent magnet.

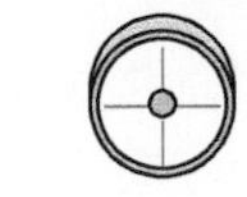

5 More wind farms are being built across the country. Give three problems with relying too much on wind power.

6 A large HEP station is currently under construction in China.

a What does HEP stand for?

b Give **three** advantages of HEP.

7 Some students have made a model of the electromagnets used in scrapyards to lift scrap cars.

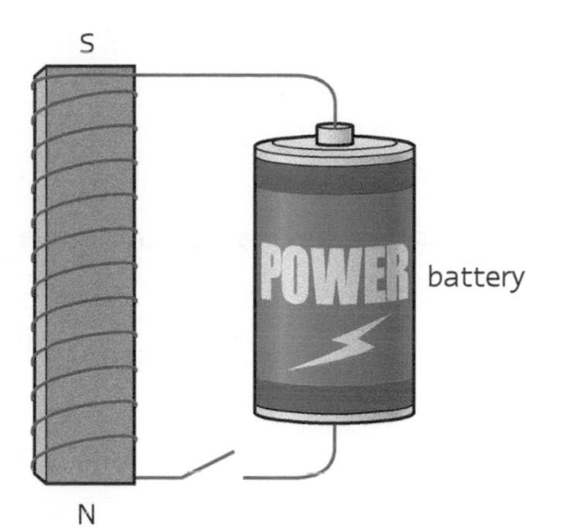

a The crane must be able to drop the cars. How can this be done? Explain your answer.

b At first the magnet they are using is not strong enough to lift the toy cars. Suggest two things that the students can do to solve this problem, using the same iron core.

8 Michael Faraday found a way of producing motion from electricity. By placing a current carrying wire in a magnetic field, the wire can be made to move. If the current is switched off, the wire no longer moves.

a Explain why the wire only moves when the current flows through the wire.

b Give **two** ways to increase the size of the movement.

c Give **two** ways to change the direction of the movement.

9 The diagram shows a simple bicycle dynamo. Study the diagram then answer the questions.

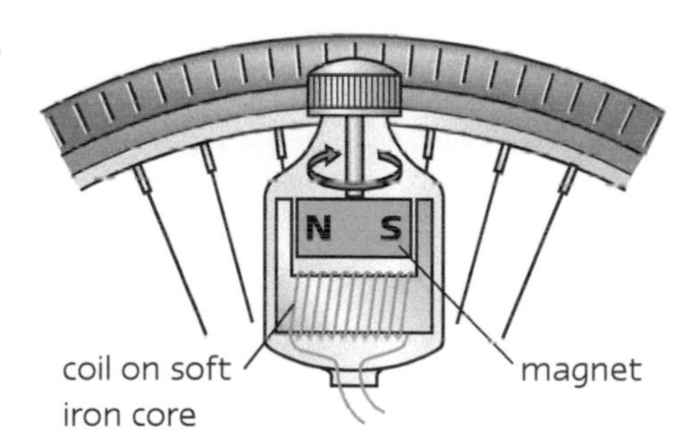

a What is the role of the soft iron core?

b Explain how the dynamo works to provide lights for a bicycle.

c What is the danger to a cyclist of using these types of lights in towns or in hilly areas?

10 You have been asked to design a power supply system for an island with 3000 inhabitants 30 minutes from the headland. Fishing and tourism are their main industries. Explain and justify the type(s) of power generation you would recommend.

Learning Checklist

4

- ☆ I know that the Earth has a magnetic field around it. page 106
- ☆ I know what type of fuels are used in power stations. page 118
- ☆ I know how power stations work. page 118
- ☆ I know how to classify some renewable and non-renewable resources. pages 122, 124

5

- ☆ I can describe the shape and direction of a magnetic field. page 104
- ☆ I can explain how magnetic materials can be magnetised using a simple domain model. page 108
- ☆ I know how to increase the strength of an electromagnet. page 110
- ☆ I can describe how electricity is generated using fuels and describe some of the possible environmental effects. page 118
- ☆ I can describe how renewable energy resources can be used to generate electricity and provide heating. page 122

6

- ☆ I can describe some of the effects of the Earth's magnetic field. page 106
- ☆ I can describe methods to demagnetise magnetic materials. page 109
- ☆ I can describe the shape of the field around an electromagnet. page 111
- ☆ I can compare the advantages and disadvantages of a range of energy resources. pages 120–125

7

- ☆ I can suggest reasons why the Earth's magnetic field may not be constant over time. page 106
- ☆ I know that magnetic effects are used in electric motors and generators. page 114
- ☆ I can describe some of the issues associated with using nuclear power. page 124

8

- ☆ I can describe some of the social, economic and environmental issues arising from our continued reliance on fossil fuels. page 120
- ☆ I can describe some of the issues associated with using nuclear power. page 125

Topic Quiz

1 Name **two** magnetic materials.
2 What happens when a magnet is placed near to a magnetic material?
3 What could happen when two magnets are brought close together?
4 Give **two** examples of the use of an electromagnet.
5 What is a magnetic domain?
6 Give **three** ways to increase the strength of an electromagnet.
7 How can a compass be used to find the shape of a magnetic field?
8 What happens to the direction of an electromagnetic field when the direction of the current in the wire is reversed?
9 What is a non-renewable energy resource?
10 Give **two** problems of transporting oil.
11 Give **two** examples of pollution caused by burning fossil fuels.
12 Name **three** renewable energy resources.

True or False

If a statement is false then rewrite it so it is correct.

1 Electromagnets stay magnetised when the current is turned off.
2 Increasing the strength of the current increases the strength of an electromagnet.
3 The Earth's magnetic field protects us from cosmic radiation.
4 The direction of the Earth's magnetic field is from geographic north to geographic south.
5 Electric motors are used to generate an electric current in a wire.
6 Nuclear power stations provide 25% of the world's electricity.
7 Only a few countries have the technology to use nuclear power.
8 Nuclear fuel is a safe fuel to use for generating electricity.
9 Nuclear power stations are cheap to build and run.
10 The explosion of Chernobyl only affected a small area for a short time.

ICT Geography/Citizenship Activities

You live in a remote area of outstanding natural beauty near the coast. You have heard that the Government is planning to build a new power station just along the coast from your home. There is a rumour that it will be one of the next generation of nuclear power stations.

The local council has persuaded a Government minister to attend a meeting of the local residents to discuss its proposals and is asking for people to come to the meeting to question the minister about the plans and what they will mean to the community.

You want to express your views and have decided to find out some information first.

- What sort of information do you think you need?
- Where do you plan to get this information from?
- What will you say at the meeting?
- How will you get your ideas across?

Noise Pollution

Noise pollution is when any unwanted sound intrudes into our lives. While noise is not the first thing that comes to mind when you think about pollution, it is becoming a bigger problem.

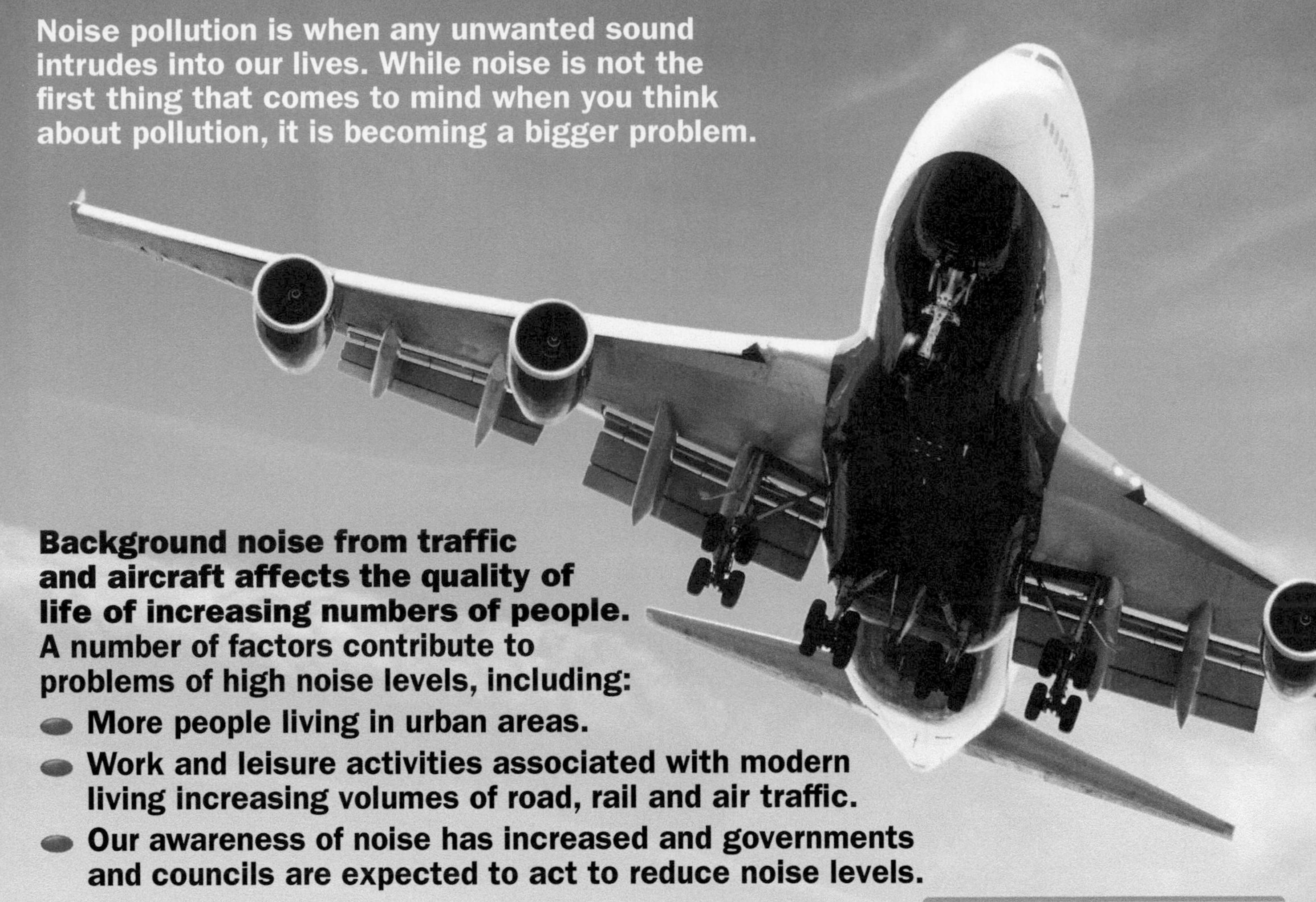

Background noise from traffic and aircraft affects the quality of life of increasing numbers of people.
A number of factors contribute to problems of high noise levels, including:

- More people living in urban areas.
- Work and leisure activities associated with modern living increasing volumes of road, rail and air traffic.
- Our awareness of noise has increased and governments and councils are expected to act to reduce noise levels.

FIGURE 1:
When an aircraft flies very fast it breaks the sound barrier and creates a loud sonic boom.

SOUND AND HEAT

BIG IDEAS

By the end of this unit you will be able to explain how energy can be transferred as heat or sound. You'll be able to describe how heat travels, you'll be able to explain the various ways in which sound can be produced and how sounds vary according to amplitude and frequency. You'll be able to use scientific models in your explanations.

FIGURE 2: A traffic jam.

One in eight people in the UK are affected by noise from aircraft. In October 2001 eight London residents living under the flight path of Heathrow took a case to the European Court of Human Rights claiming that night flights violated their human rights by denying them a normal night's sleep. However there is a planned major expansion in aviation with new runways at several airports.

FIGURE 3: Noise levels in the countryside have increased because of industrial farming.

This can only increase the noise problems experienced by local communities.

But it's not just aviation that is to blame for excessive noise. The European Commission estimates that 90 per cent of background noise comes from road traffic.

In its Green Paper on the development of a future policy on noise in 1996, the European Commission estimated that between 17 and 22 per cent of the population of the EU suffered from background noise levels that were unacceptably high.

Closer to home, a survey carried out for Campaign for Rural England looking at rural tranquillity has found that the countryside is becoming more affected by urban noise. In the early 1960s 70 per cent of England could be described as tranquil or free from intrusive noise. Thirty years later this had fallen to 56 per cent.

What do you know?

1 State **five** things that you know about sound.

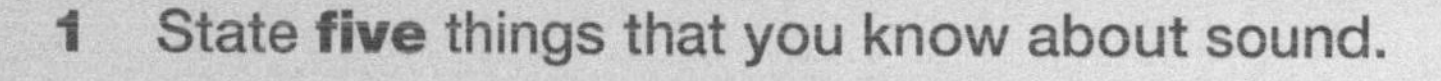

2 What is noise pollution?

3 How do we hear sound?

4 Give **three** examples of noise pollution.

5 Give **three** examples of noise pollution caused by leisure activities.

6 Where does most of our background noise come from?

7 How many people suffer from noise pollution?

8 How do you think the countryside is affected by urban noise?

9 Why do you think there is a need for more runways to be built?

10 Why do you think the people living near Heathrow had to take their case to the European Court of Human Rights?

11 Should military aircraft have to reduce speed in built-up areas? Explain your answer.

12 What could governments and councils do to reduce noise pollution? What issues will they need to consider?

What is sound?

BIG IDEAS

You are learning to:

- Understand what causes sound
- Understand how to change the loudness and the **pitch** of a sound HSW

Good vibrations

When an object vibrates it makes a **sound**. Different objects make different sounds when they **vibrate**. In many cases the vibrations may be difficult to see or they may be invisible.

The different musical **instruments** in an orchestra make sound in many ways. Most orchestras contain stringed instruments, wind instruments and brass instruments. By working together the orchestra can produce a range of sounds which allows them to perform complicated pieces of music.

FIGURE 1: An orchestra playing together to produce a range of sounds.

1. Name **five** instruments in the orchestra.

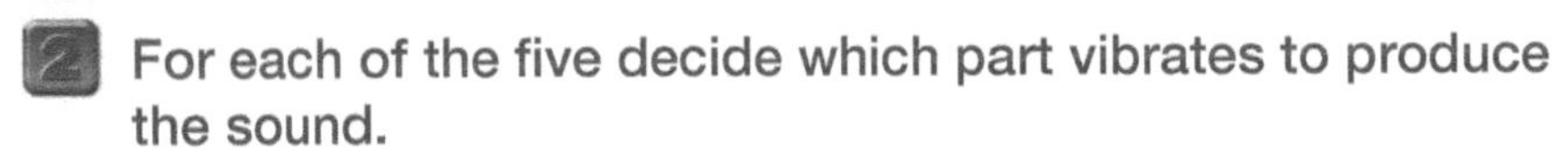

2. For each of the five decide which part vibrates to produce the sound.

Did You Know...?

The name for the piano is a shortened version of the Italian *pianoforte*: its note can be soft (*piano*) or loud (*forte*) depending on how hard the key is pressed.

Getting louder

Most musical instruments can make sounds which can be **loud** or **soft**. For stringed instruments the sound can be made louder by plucking or hitting the string harder. Wind and brass instruments can make louder sounds when the musician blows harder. This increases the energy of the vibrations. So loud sounds have more sound energy than soft or quiet sounds.

3. Describe how a trumpet player can make his or her instrument sound louder.
4. Explain why the job of the conductor is so important to the orchestra.
5. Can you explain how the sound reaches the audience?

Science in Practice

Bells and whistles

Most objects can be made to vibrate and can then make a sound. The type of sound that an object makes depends on which part vibrates. Your teacher will provide you with a range of different objects and some musical instruments.

FIGURE 2: Bells

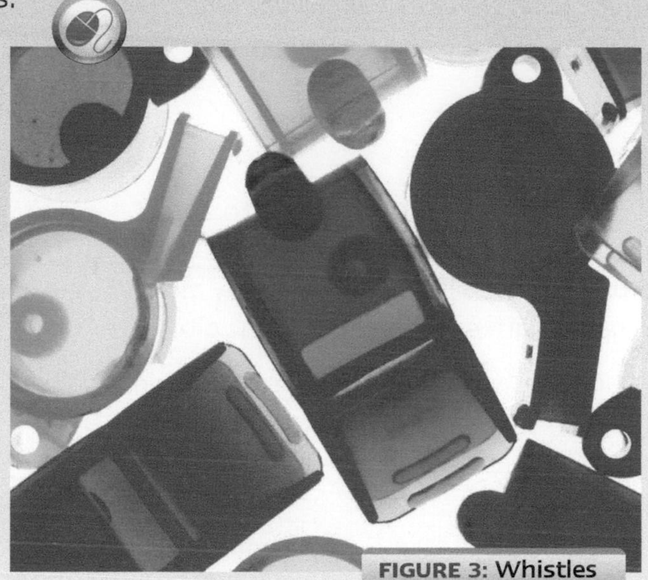

FIGURE 3: Whistles

Look at each object carefully and try to decide how it makes a sound. What part of the object is vibrating each time? For each of the objects, think about:

- What vibrates to make the sound?
- Can you see the vibrations?
- How can you make the sound louder?
- How can you make the sound have a higher pitch each time?

Questions

1. Write a paragraph to describe what you have found out. Use some or all of the key words for the topic to help you.

2. Why is there a limit to how loud you can make the sounds?
3. Why do loud sounds have more energy than quiet sounds?
4. How could you change the pitch of the sound you have made?

Describing sounds

BIG IDEAS

You are learning to:

- Understand how sound waves can be compared using an oscilloscope (HSW)
- Explain the meaning of frequency, wavelength and amplitude
- Describe sound waves

Seeing the sound

How would you describe a sound to a deaf person? You could tell them how loud it is or what **pitch** the sound is. Some instruments such as bells only make one note, which can be a bit dull so bell ringers never play alone. Some instruments such as pianos and guitars can make many sounds in different pitches.

When objects vibrate faster they make higher pitched sounds. The pitch of a note is called its **frequency (f)** and is measured in **Hertz (Hz)**.

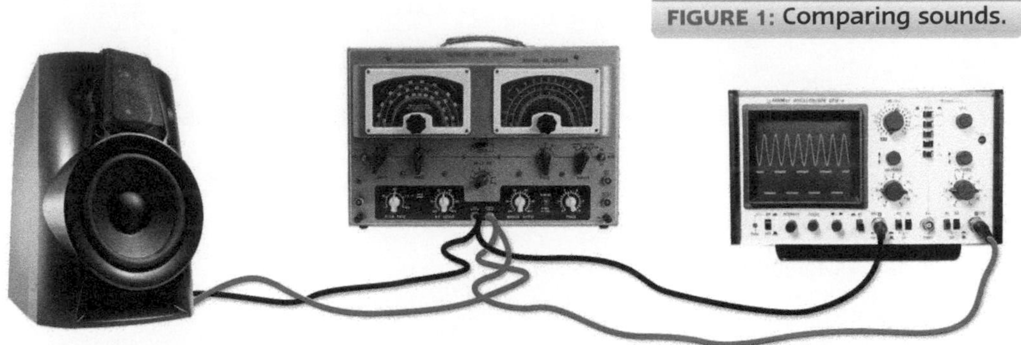

FIGURE 1: Comparing sounds.

Connecting a loudspeaker and an **oscilloscope** to a **signal generator**, as shown in Figure 1, lets us compare sounds. The signal generator produces sounds at different frequencies. The loudspeaker enables us to hear and compare the sounds. The oscilloscope displays a picture of the sound as a **wave pattern** on a screen.

At low frequency a low or **bass** note is produced. As frequency increases the waves on the screen get closer together and have a shorter **wavelength**.

FIGURE 2: A low frequency sound.

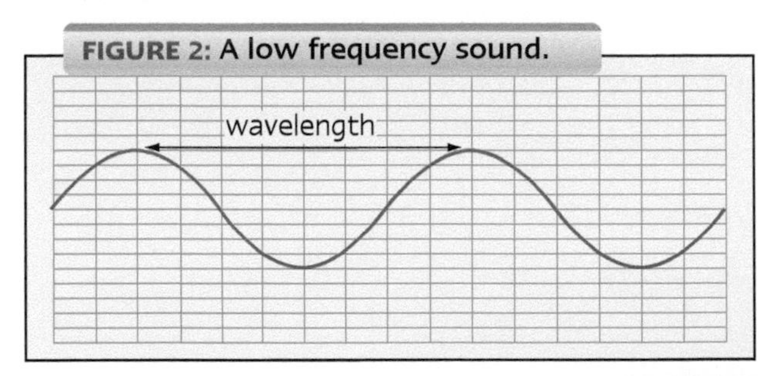

FIGURE 3: A high frequency sound.

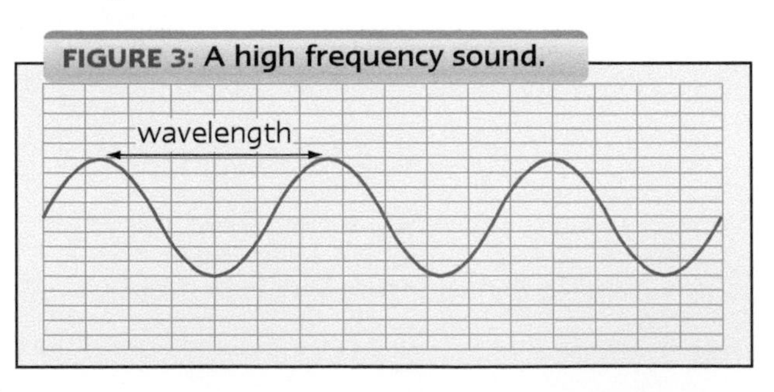

Loud sounds have more energy than quiet sounds. A loud sound has a taller wave pattern than a quiet sound of the same frequency.

FIGURE 4: A quiet sound.

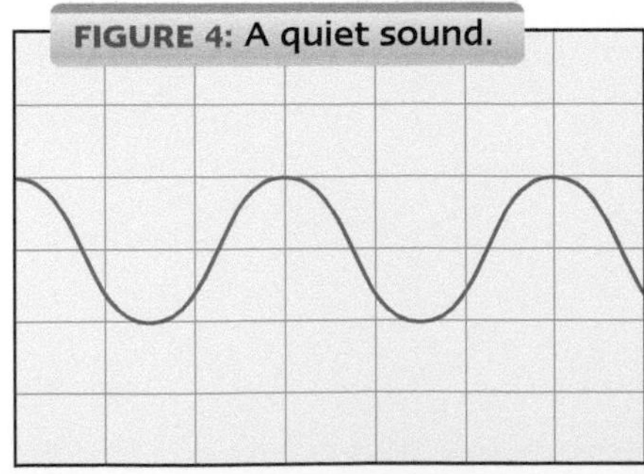

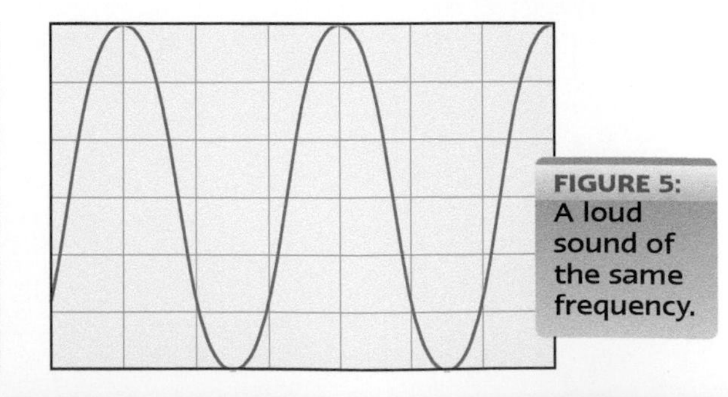

FIGURE 5: A loud sound of the same frequency.

... amplitude ... bass ... frequency ... Hertz (Hz) ... oscilloscope

1 Which of Figures 1, 2, 3 or 4 shows the wave with the lowest pitch? Explain your answer.

2 Which has the shortest wavelength? Explain your answer.

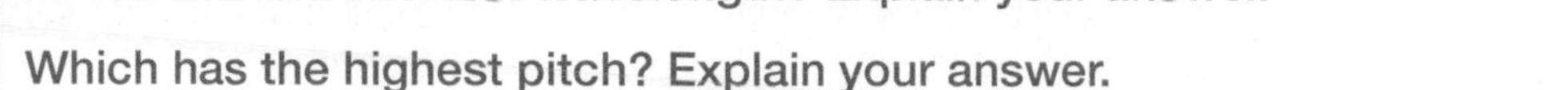

3 Which has the highest pitch? Explain your answer.

FIGURE 6: Sound waves showing amplitudes.

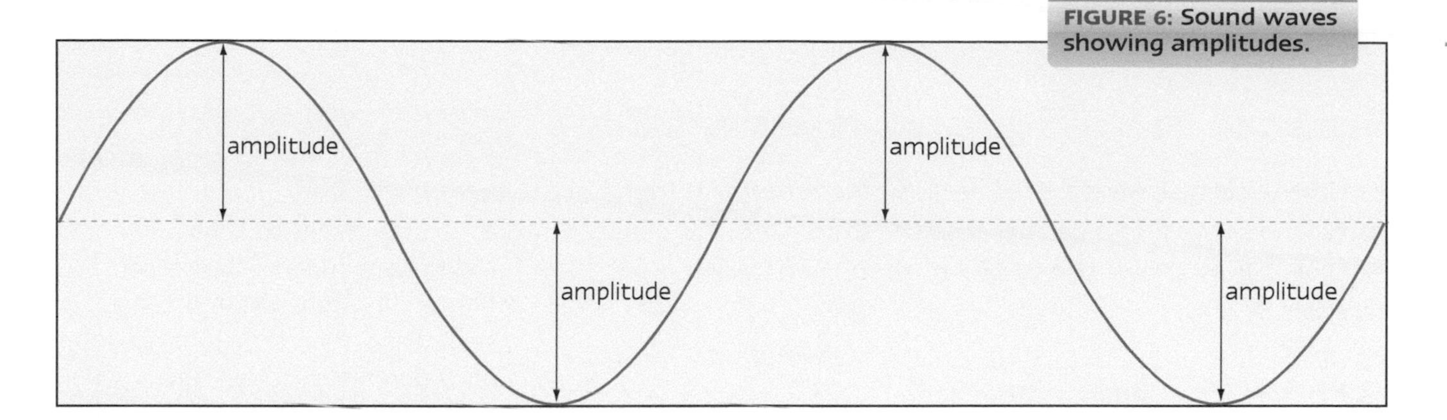

Describing the sound wave

The amount of energy that a sound wave has is called its **amplitude.** Because loud sounds have more energy they have a bigger amplitude than quiet sounds.

4 Sketch a wave shape for the following **three** sounds.
a A quiet high frequency sound
b A loud low frequency sound
c A loud high frequency sound

Sound energy

Because a sound is produced by an object vibrating back and forth the amplitude is the maximum distance the particle moves from its starting position.

The waveform in Figure 4 has a smaller amplitude, but the same frequency, as the waveform in Figure 5. The only difference between the sounds is that the first one is quieter, as it has less energy.

5 Describe how pitch and frequency are related.

6 Explain the difference between frequency, wavelength and amplitude.

Watch Out!

Remember that the amplitude shows how far the particle has moved from its starting position in either a forwards or a backwards direction. So the amplitude is half the total height of the wave from its trough to its crest.

Speed of sound

BIG IDEAS

You are learning to:

- Recognise that the speed of sound is different in different materials
- Describe how echoes can be used 
- Calculate the speed of sound

Sound in different materials

Sound travels fastest in a solid because the material is more **dense** and slowest in a gas because the material is less dense. The table below shows the **speed** of sound in different materials.

Material	Speed (m/s)
air	330
water	1500
brick	4300
iron	5000

1. What does 'density' mean?
2. How far does sound travel through air in 3 seconds?
3. How far does sound travel through water in 2 seconds?

Did You Know...?

Native Americans used to listen to railway tracks to know when trains were coming. Resistance fighters in Afghanistan use the same trick when they want to blow up enemy troop trains because it gives them more time to carry out the attacks effectively.

Thunderstorms

In a thunderstorm you see a lightning strike before you hear a thunder clap. If the storm is distant it takes longer to hear the thunder. If the storm is very distant the thunder may sound very faint. The further away you are the more the sound energy spreads out (**dissipates**) before it gets to you and the harder it is to hear it.

4. Which travels slowest, light or sound? Explain how you know this.
5. Sam hears thunder 5 seconds after he sees lightning. How far away is the storm?
6. How can Sam tell if the storm is getting closer or moving away?

FIGURE 1: In a thunderstorm do you hear thunder before or after you see lightning?

... dense ... dissipate ... echo ... echo-sounding

Echoes

Echoes are caused when sound waves bounce back (are **reflected**) from some surfaces. Harder surfaces reflect sound more than soft surfaces.

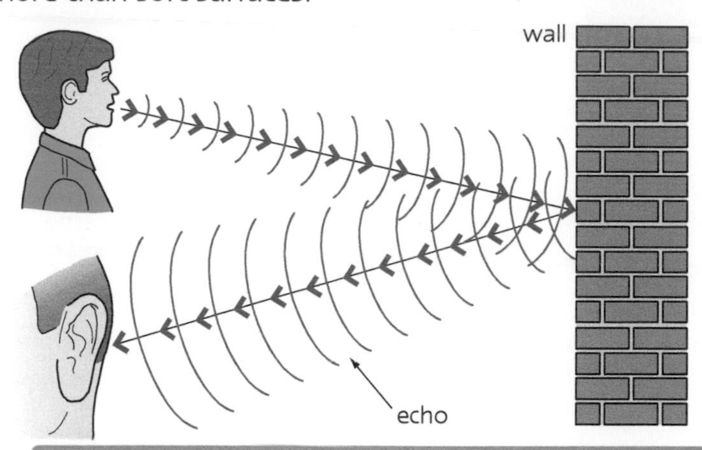

FIGURE 2: How echoes are made. Would there be an echo if there was a line of trees in place of the wall?

> **Watch Out!**
>
> In echo-sounding, the sound waves need to travel to the seabed and back. So the time it takes to reach the bottom is half the time it takes for the whole journey. Remember this if you are asked to find the depth of the water!

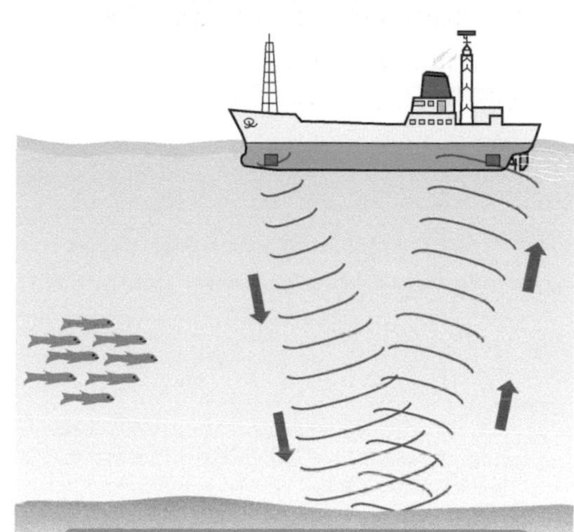

FIGURE 3: How fishing boats use sonar to locate shoals of fish.

Fishermen use **echo-sounding** or **sonar** to locate shoals of fish. Sound waves are sent out from the boat and reflect from the seabed. If a shoal of fish is passing under the boat then the sound waves are reflected by the fish and the echo gets back to the boat more quickly because the sound has less distance to travel. Battleships use echo-location to detect submarines in the same way.

7. Using sonar an echo is heard on a boat after 2 seconds. How deep is the sea?
8. A shoal of fish passes 750 metres below a boat using sonar. How long does it take to hear the echo from the fish?
9. Explain how a battleship knows at what depth to set its depth charges when it is hunting a submarine.

Calculating speed

The speed of sound is calculated using the equation:

$$\text{speed of sound (m/s)} = \frac{\text{distance travelled (metres)}}{\text{time taken (seconds)}}$$

> **Exam Tip!**
>
> For calculations using speed remember to use the correct units:
> - speed in metres per second (m/s)
> - distance travelled in metres (m)
> - time taken in seconds (s).

10. Calculate the speed of sound in a metal railway track that is 1200 metres long if the sound is heard at one end, 0.5 seconds after it is made at the other end.
11. How far away is a quarry if the sound of the blasting is heard 3.5 seconds after the explosion?
12. How long does it take for a sailor to hear the echo from a submarine if the submarine is 3500 metres away from the ship?

Sound waves

BIG IDEAS

You are learning to:

- Explain what a sound wave is
- Recognise the difference between a compression and a rarefaction
- Understand how a loudspeaker produces sound

Making waves

Pushing one end of a spring back and forth makes some parts squash together and some parts pull apart. A pulse, or **wave**, of **energy** passes down the spring. This is called a **push wave**. The energy passes through the spring.

longitudinal wave

source moves left and right

coils move left and right

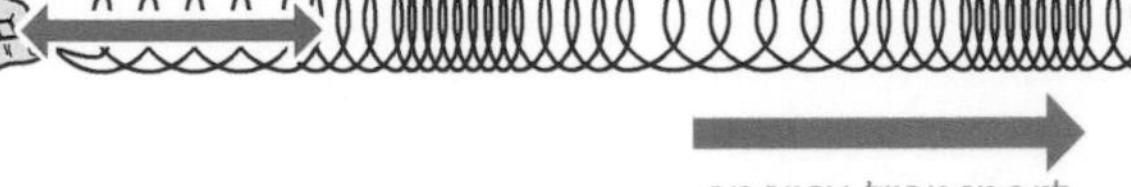

energy transport

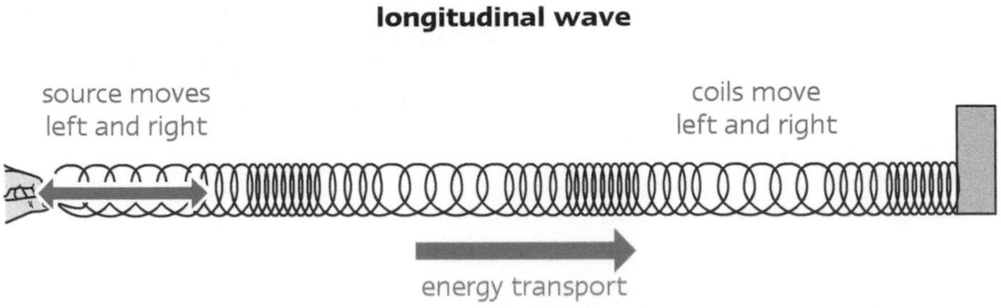

FIGURE 1: Making a push wave in a spring. What happens to the energy?

1. Describe how to make a wave in a spring.
2. What is the wave shown in Figure 2 called?

Compressions and rarefactions

When a wave is passing along a spring at any given point, its coils are pushed close together producing a high-pressure area or **compression** and then they are pulled apart producing a low-pressure area or **rarefaction**.

FIGURE 2: What happens when a spring is pushed and pulled?

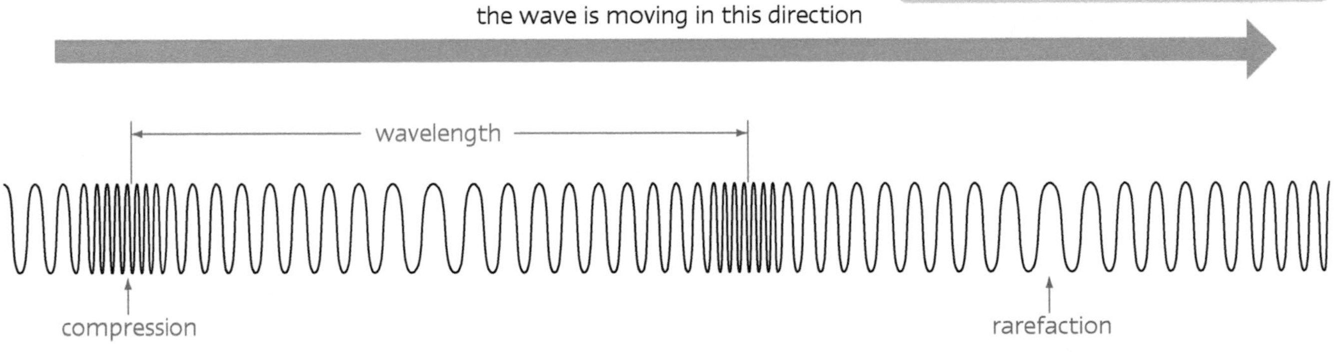

the wave is moving in this direction

wavelength

compression

rarefaction

During a wave, energy passes through the spring in the same direction as the backwards and forwards vibrations. This type of wave is called a **longitudinal** wave. The **particles** in the spring vibrate backwards and forwards from their starting positions. They do not move along the spring. So the particles in the coils **transmit** the energy through them.

Air behaves in a similar way. When a person speaks or shouts their vocal chords vibrate and this sets up sound waves that are transmitted by the particles in the air.

... compression ... dissipated ... energy ... longitudinal ... particle

3 Explain what 'compression' and 'rarefaction' mean.

4 What is a longitudinal wave?

5 What happens to the particles in a spring when a longitudinal wave passes along it?

Sound travels

When a loudspeaker produces a sound a cone inside the speaker vibrates forwards and backwards very quickly.

- When a loudspeaker cone moves forwards the air particles are pushed close together, the air pressure rises and a compression is produced. The sound energy is passed on when the air particles move and bump into nearby air particles. The compression moves outwards through the air.
- When a loudspeaker cone moves backwards the air molecules are pulled apart, the air pressure falls and a rarefaction is produced.

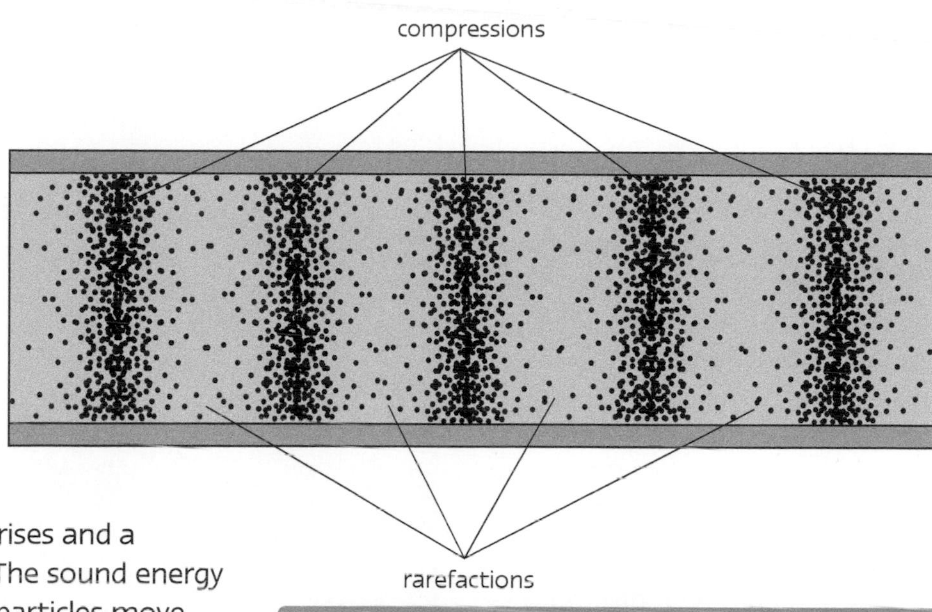

FIGURE 3: This is what compressions and rarefactions look like. The energy of the wave is transmitted through the air. What carries the energy in the air?

Every time the cycle repeats a new sound wave is produced. This continual pushing and pulling of the particles in the air creates a series of waves that **transfers** the energy from the loudspeaker outwards as a sound wave.

As a person moves away from a sound source the sound energy becomes **dissipated** (spread out) and it is harder to hear.

6 Describe how the sound from a loudspeaker travels through the air.

7 Explain using ideas about energy and particles why it is quieter at the back of a concert hall than at the front.

Did You Know...?

Megaphones are used to amplify sound produced by voices so that the sound travels further.

Did You Know...?

In the deep ocean the sperm whale uses sound to stun or kill its prey. It sends out giant grunts, immensely powerful bursts of sound that can disable nearby fish, squid and other victims.

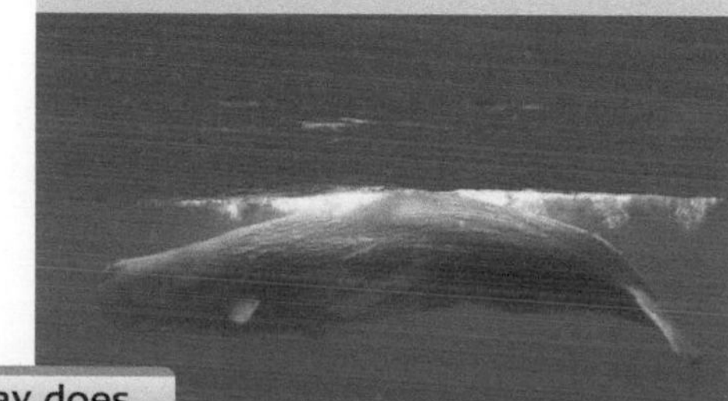

FIGURE 4: In what way does a sperm whale use sound?

Sounds in solids, liquids and gases

BIG IDEAS

You are learning to:
- Understand how sound travels through materials
- Use a model to explain how particles help transmit sounds

At a rock concert

Most of the sound that we **hear** travels through the air (a **gas**) to our ears. Sound can also travel through **solids** and **liquids** at different **speeds**.

1 What does most of the sound that we hear travel through?

2 Name **two** other materials that sound can travel through.

FIGURE 1: How do you hear the sound from a rock concert? Where is the best place to stand to hear the band?

Sound in a vacuum

Because outer space is a **vacuum** there is no sound. So astronauts can only talk to each other through radio links.

3 What is a vacuum?

How Science Works

We can model this by using a gas jar and a bell. An electric bell is placed in a gas jar and then switched on. The bell is clearly heard as the sound travels through the air in the jar and the glass surrounding it. A vacuum pump is used to remove the air from the gas jar and the sound gets quieter and quieter until it is no longer heard.

FIGURE 2: Demonstrating that sound cannot travel in a vacuum. Explain why the sound from a bell in a gas jar gets quieter if the air is removed.

HSW

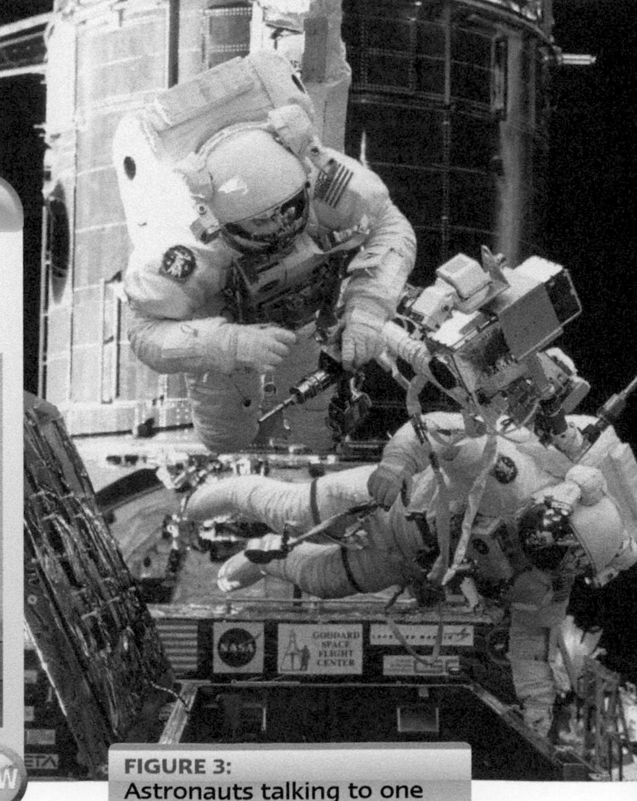

FIGURE 3: Astronauts talking to one another through radio links.

Sound and particles

In a gas such as air the **particles** are not joined together and move freely.

A **sound wave** passes through the air because the moving particles bump into their neighbours. When they collide the particles pass the **vibration** energy on and the sound wave moves through the air.

All materials are made of particles. Sound travels at different speeds in different materials because of the different ways that the particles are arranged.

- In a gas — the particles are far apart and sound travels slowly because the particles do not collide very often.
- In a liquid — the particles in a liquid are more closely packed and sound travels more quickly because the particles collide with each other more frequently.
- In a solid — the particles are packed closely together. They cannot move but they can vibrate around fixed positions and sound passes very quickly because the particles collide easily with their neighbours.

So putting our ear to the ground does enable us to hear things quicker but it is not very comfortable!

4 Use the particle theory to explain how sound travels through a solid.

5 Why does sound travel slowest through gases?

6 Suggest why sound does not travel through foam.

FIGURE 4: The particle theory explains how sound travels in a gas, a liquid and a solid.

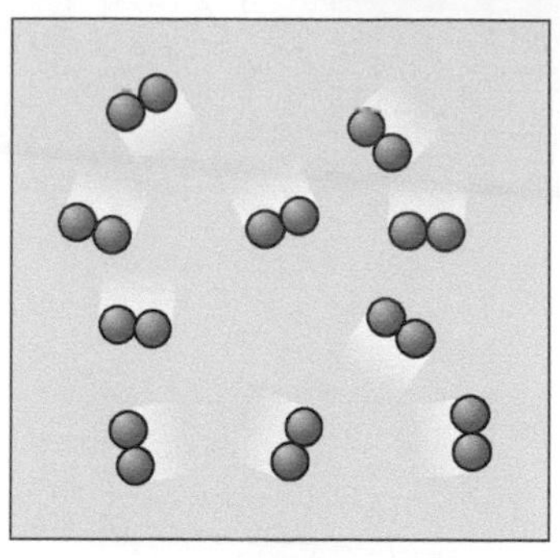

gas

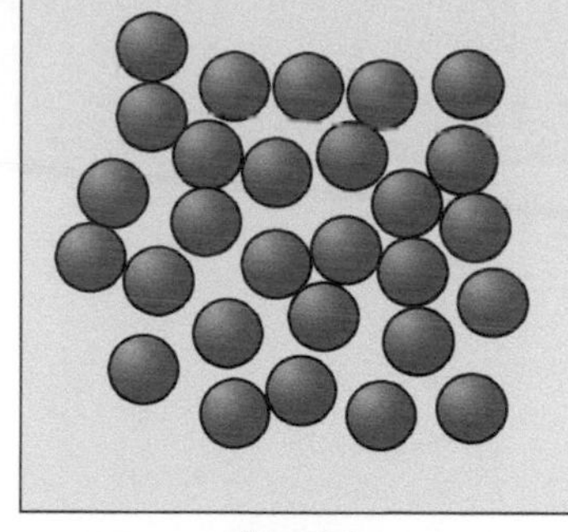

liquid

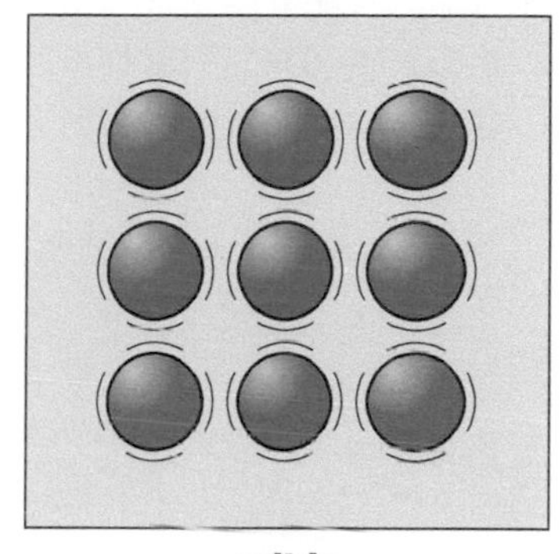

solids

vacuum

How Science Works

It is sometimes useful to be able to stop sound passing through materials.
Recording studios are lined with special materials made of soft foam moulded into shapes. The foam absorbs the sound energy so there are no sound reflections or echoes to interfere with a recording.

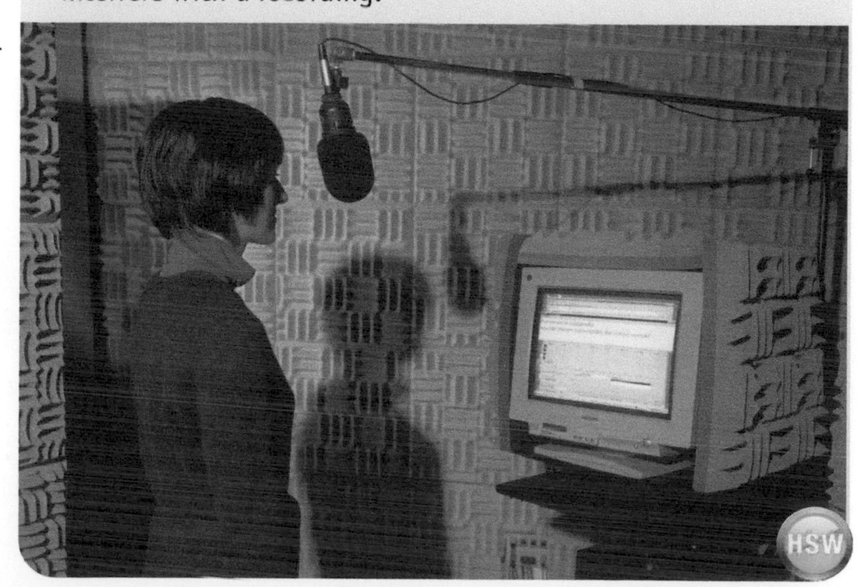

... *sound wave* ... *speed* ... *vacuum* ... *vibration*

Ultrasonic sounds

BIG IDEAS

You are learning to:
- Describe the hearing range in humans
- Explain what ultrasonic sound is
- Recognise the uses of ultrasonic sound

Human hearing range

As the **frequency** of a sound increases it becomes higher **pitched** until our ears can no longer hear it. Our ears only let us hear sounds of certain frequencies. The **range** of frequencies we can hear is called our **hearing range**. In humans the hearing range is normally between about 20 Hz and 20 000 Hz. The highest frequency we can hear is called the **upper threshold** of our hearing. As we get older the upper threshold gets lower.

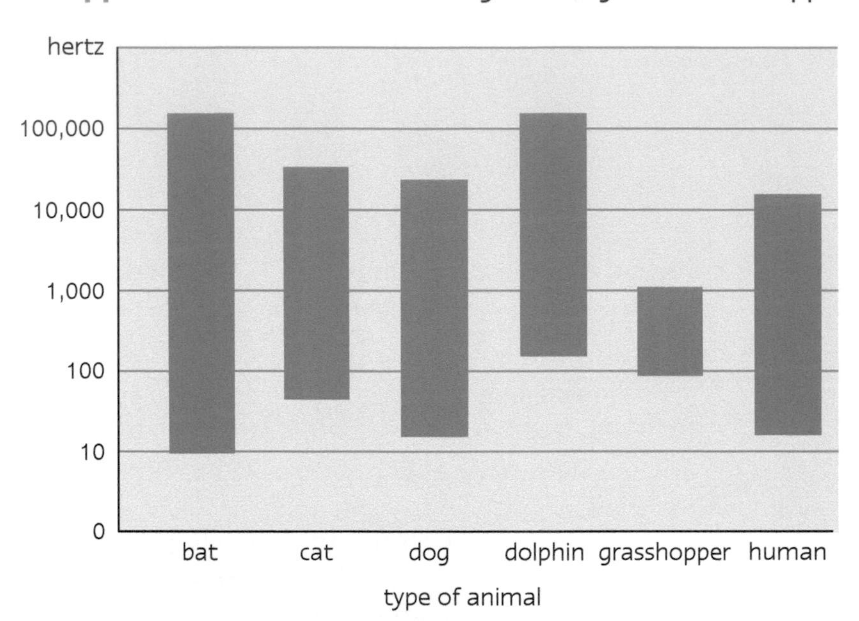

FIGURE 1: Which animal on the graph has the biggest hearing range?

Did You Know...?

Some animals can hear sounds whose frequencies are above our upper threshold.
An example is dogs. Special dog whistles have been made that are silent to humans.

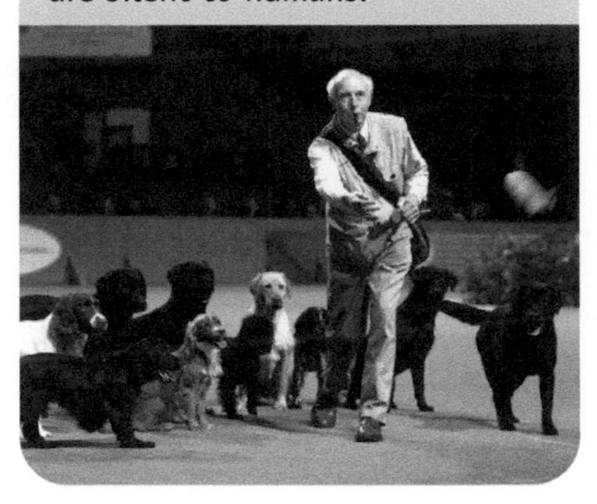

1. What does 'frequency' mean?
2. What is the normal human hearing range?
3. Why do you think the upper threshold of hearing gets lower as people get older?

Using ultrasonic sound waves

Sounds with a frequency above the human hearing threshold are called **ultrasonic sounds**.

Because ultrasonic sounds (**ultrasound**) are at such a high frequency it is easier to **direct** the sound waves. Ultrasound has many uses.

... direct ... echolocation ... frequency ... hearing range ... pitch

How Science Works

- In medicine:
  - ultrasound is used to scan organs such as the heart, brain or liver without the need for surgery. The most common use of ultrasound is in a pregnancy scan. Ultrasonic waves are used to build up an image of the developing baby in the mother's womb. This enables a doctor to check the progress of a pregnancy.

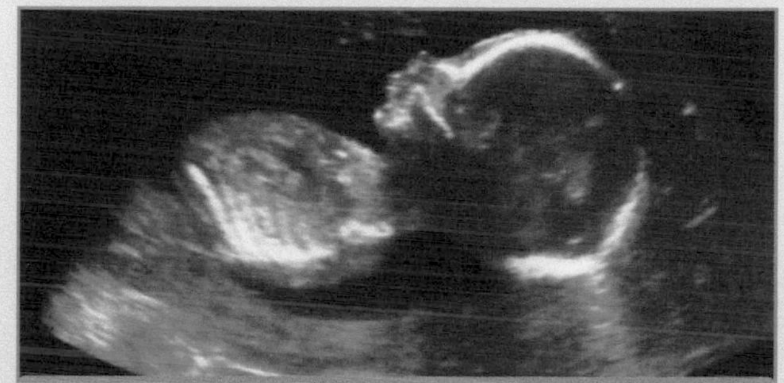

FIGURE 2: Ultrasound scan of a human foetus made at 20 weeks. The ultrasound echoes are recorded and build up a picture of the foetus.

  - kidney stones are broken up without the need for surgery using ultrasound. The high-frequency wave vibrations are directed at the stones which causes them to vibrate and break up. They are then passed painlessly from the body.
- In dentistry:
  - ultrasound is used to remove plaque from teeth. The high-frequency wave vibrations cause the plaque to vibrate so much that it breaks up and falls off.

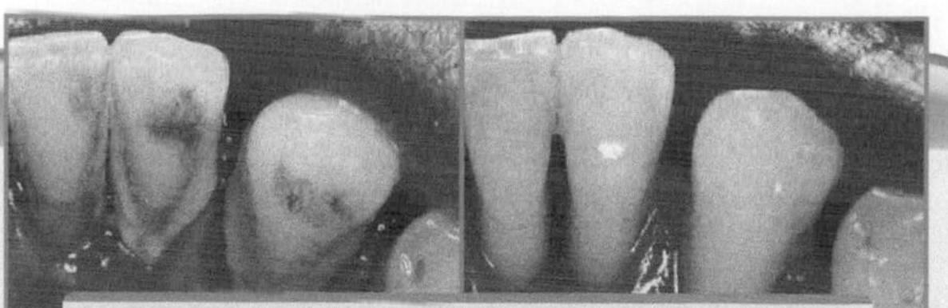

FIGURE 3: 'Shiny and white'! Ultrasound was used to clean the brown-coloured plaque off these teeth.

- Cleaning:
  - a similar effect is used to clean many different types of objects, especially surgical instruments. The objects are immersed in a liquid and ultrasonic waves are directed through it. The high-frequency waves create a 'scrubbing' action in the fluid and any unwanted material is removed from the objects.
  - ultrasound is also used to clean delicate machinery without the need to dismantle it.
- Cracks:
  - because ultrasound reflects off cracks in metals it can be used to check for faults in aircraft and in underground pipes.

FIGURE 4: These gas cylinders are being checked for defects such as cracks using ultrasound.

HSW

4 **a** Explain what 'ultrasonic' means?
b Describe **three** uses of ultrasound.

5 Suggest why medical instruments need to be very clean.

6 What advantages and disadvantages are there in using ultrasound instead of X-rays to check the progress of a pregnancy?

Did You Know...?

Bats and dolphins use **echolocation** to communicate, to hunt and to help them navigate. They send out ultrasonic 'clicks' and can tell from the type of echo they receive back whether the object is prey or an object that needs avoiding. A bat can feed successfully in total darkness.

The ear and hearing

BIG IDEAS

You are learning to:

- Describe the structure of the ear
- Understand what the different parts of the ear do
- Explain how our ears enable us to hear

Structure of the ear

Everyone's ears look slightly different on the outside but they all work in the same way. Our ears are not just the fleshy bits that we can see. When sound waves travel towards a person they enter his or her ears. Once inside the sound waves cause the **eardrums** to **vibrate** which enables the person to hear the sound.

The ear is divided into three sections:

- the **outer ear**
- the **middle ear**
- the **inner ear**.

The outer ear is seen on the outside of the head. The parts of the ear that enable us to hear sounds are very delicate. They are in the middle ear and inner ear inside the head and are protected by the bony skull.

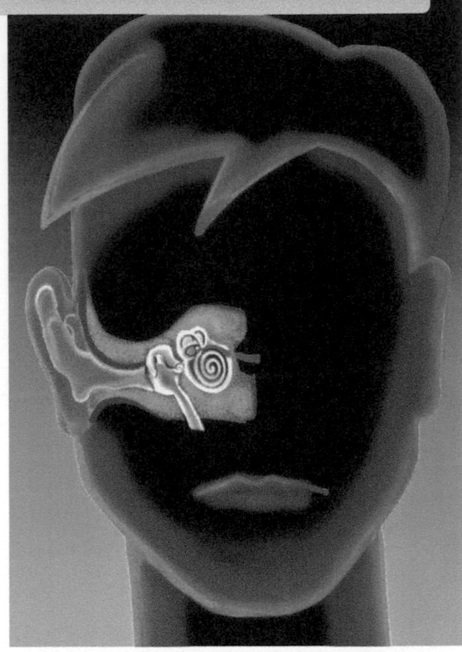

FIGURE 1: This is where our ears are inside our heads. Why do you think most of the ear structure is inside the skull?

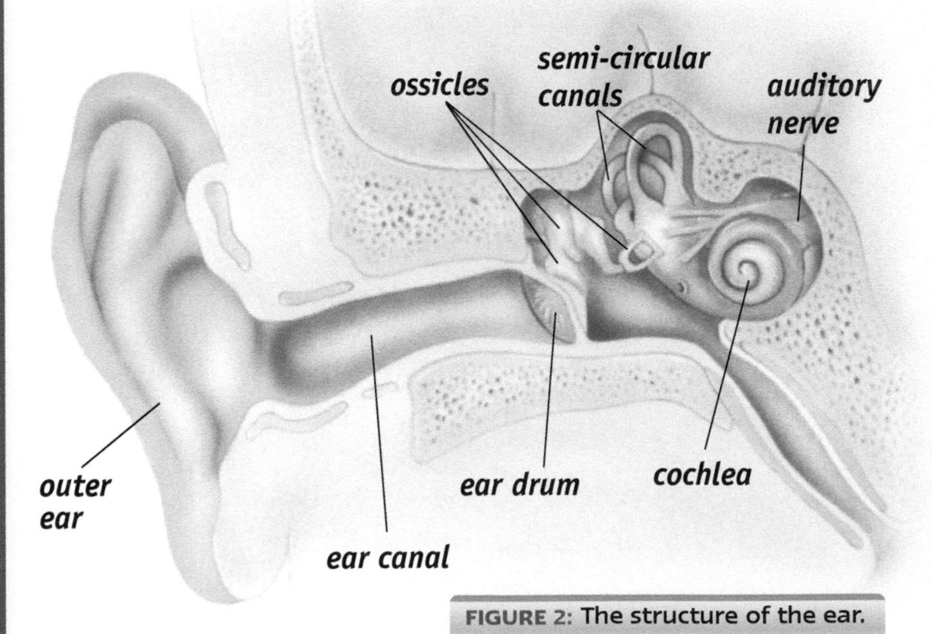

FIGURE 2: The structure of the ear.

1. What does the eardrum do so that we can hear sounds?
2. Where in the body are the important parts of the ear found?

Did You Know...?

The three ossicle bones in your ear are the smallest bones in your body.

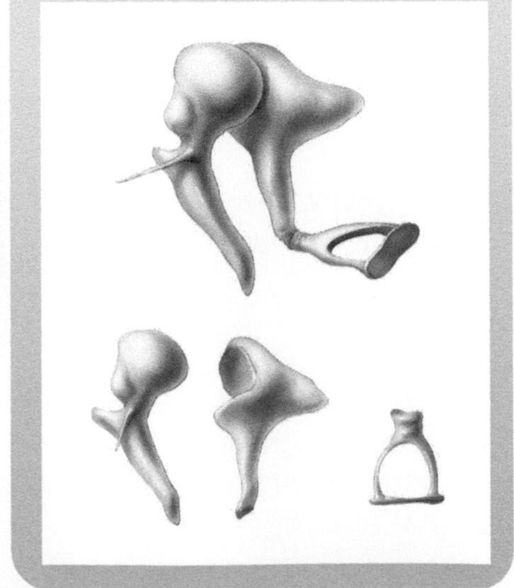

... auditory nerve ... adapt ... brain ... eardrum ... inner ear

How do ears work?

Humans' ears are **adapted** to hear a range of sounds of frequencies between about 20 Hz and 20 000 Hz. The ear works in the same way to hear all the frequencies within this range.

The ear transfers sound energy in the air around us into electrical **nerve impulses**. These impulses travel along the **auditory nerve** to the brain. The **brain** then makes sense of (interprets) the signals to enable a person to hear.

Did You Know...?

The ears are also used to help a person to balance. The semi-circular canals in the inner ear are filled with fluid. This fluid moves when the head is moved. The movement is detected by nerve cells that send a nerve impulse to the brain.
Ear infections can sometimes upset a person's balance as well as his or her hearing.

A vibrating object such as a trumpet produces sound waves.

The sound waves travel through the air to the ear.

The vibrating air enters the ear and is funnelled into the ear canal.

The ear canal seen using a very powerful scanning electron microscope.

The vibrating air sets up vibrations in the eardrum.

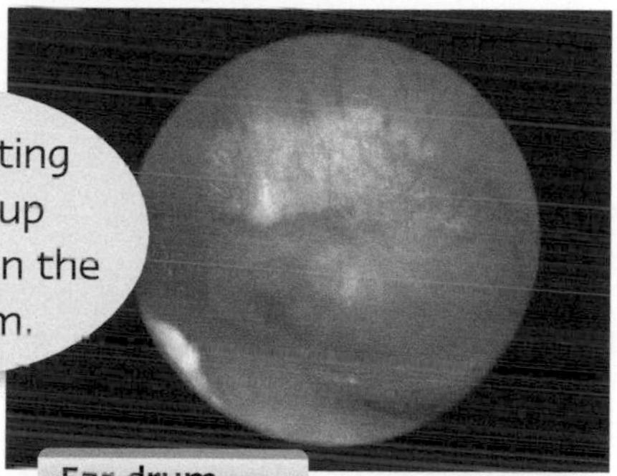

Ear drum.

The eardrum transmits the vibrations to the small bones (or ossicles) causing them to start to vibrate.

The vibrations in the ossicles cause the cochlea to start to vibrate.

The cochlea.

Sensory hair cells seen using a scanning electron microscope.

The vibrations in the cochlea are detected by hair cells lining the cochlea.

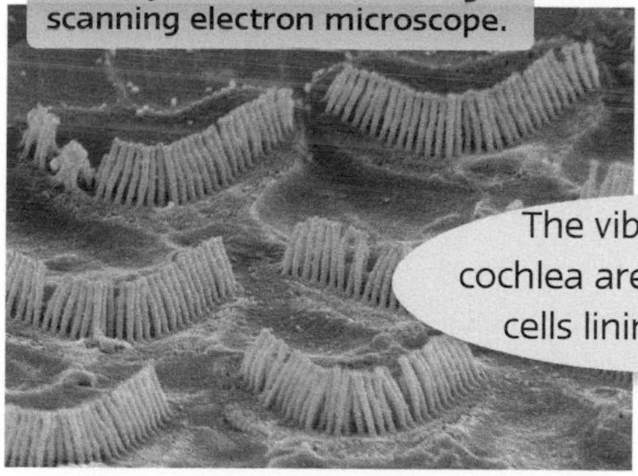

The vibrations are detected by the auditory nerve that sends a nerve impulse to the brain.

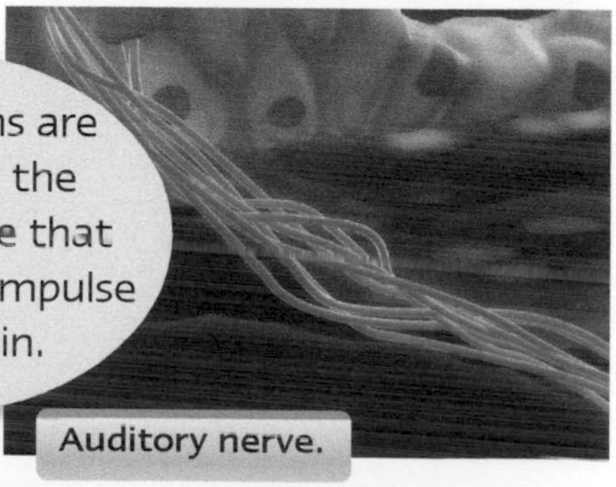

Auditory nerve.

3. Using a flowchart describe the stages involved in hearing a sound.
4. What other function does the ear perform?
5. What advantages are there in having two ears.

... middle ear ... nerve impulse ... outer ear ... vibrate

Damaging our hearing

BIG IDEAS

You are learning to:

- Understand what we can do to protect ourselves from hearing damage 
- Recognise what precautions employers need to take to protect their workers from hearing damage

How ears are damaged

Most **deafness** happens because the ear is damaged in some way.

Possible causes of damage include the following.

- The ear canal can become blocked with **wax**. This can be cleaned out.
- Very loud sounds can tear or rupture the eardrum. Usually the eardrum heals itself, but this can take a long time.
- The eardrum may be damaged by an infection.
- The small bones in the ear can become stuck together. This may need an operation to correct.
- The middle ear can become infected. This can usually be cured by antibiotics.
- The cochlea may be damaged by loud noises. There is currently no cure for this.

FIGURE 1: Why do you think this man is wearing ear defenders?

In general deafness is caused by:

- sound vibrations not being transmitted effectively in the ear
- nerve cells in the ear deteriorating. This is why older people often cannot hear as well as young people.

People who find it hard to hear sometimes use a hearing aid (page 148). 

 1 How can loud sounds harm us? 

2 Give **three** causes of deafness.

3 How can each of the causes you gave in your answer to question 2 be cured?

Measuring 'loudness'

The loudness of any sound is measured in **decibels** (**dB**). As sounds get louder they have more **energy** and their decibel level rises. The quietest sound we can hear is 0 dB.

In some cases it is very difficult to reduce the amount of noise and it then becomes important to protect our ears.

4 Give **two** examples of where it may be difficult to reduce noise levels. Why is this?

5 Estimate the loudness of:
a a whispered conversation
b a waterfall
c a train passing through a station.
For each one say how you decided on your answer.

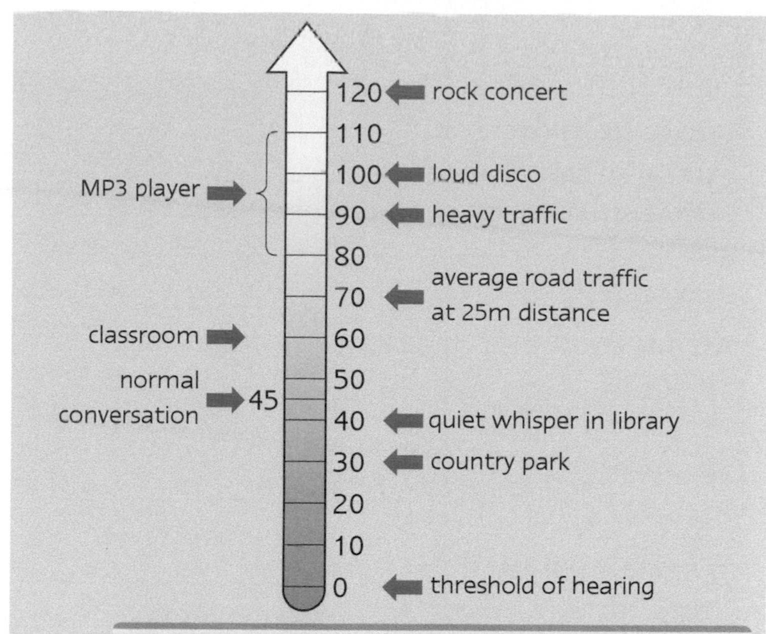

FIGURE 2: Graph to show the loudness of different sounds. In the decibel scale, each gap of 10 dB means a ten-fold increase in energy. For example 50 dB has ten times more sound than 40 dB.

Protecting hearing

To prevent a person's hearing from becoming temporarily or permanently damaged the following steps can be taken.

- Turn down the volume on personal stereos, televisions and MP3 players.
- Wear **ear defenders** when working in noisy places or with loud machinery.
- Avoid shouting.
- Use sound-proofing in factories and offices.

The government has recently passed new laws to limit noise levels in the workplace.

6 List **three** ways to reduce the risk of hearing damage.

7 Explain why you think the government has passed new hearing laws. Can you think of any problems with laws that reduce noise output?

Noise in the workplace

Employers need to help their workers by:

- reducing the amount of time they are exposed to loud noise
- considering other ways of working to reduce their exposure to loud noise
- wherever possible increase the distance between workers and noisy operations
- providing hearing protection.

8 Why are people in the construction industry at greater risk of hearing damage than people working in offices? Explain your answer.

9 What are the benefits to employers of reducing noise levels in the workplace?

Did You Know...?

Government research shows that over 2.2 million workers are exposed to noise levels above 80 dB, including over 1 million exposed to levels above 85 dB and 450 000 exposed to levels above 90 dB. Many of these workers work in the construction and house-building industries.

HSW New technology is music to deaf ears

For many years the only help for deaf people was an 'ear trumpet' that worked by making sounds louder as they passed from the horn of the trumpet into the ear. Then, as scientists understood the ear and how it works better, they developed cochlea implants. These tiny devices are implanted in the part of the ear called the cochlea and help a person to hear sounds.

In June 2007 scientists at the American University of Michigan announced what they hope will be a breakthrough in the treatment of deafness. They have developed a new type of implant that attaches to the main nerve that carries the nerve impulses from the ear to be interpreted by the brain. So far their device has only been tested on animals but the scientists hope that it will be even more effective than cochlea implants. Their device works better over a wider range of frequencies and this should let users hear a greater range of sounds.

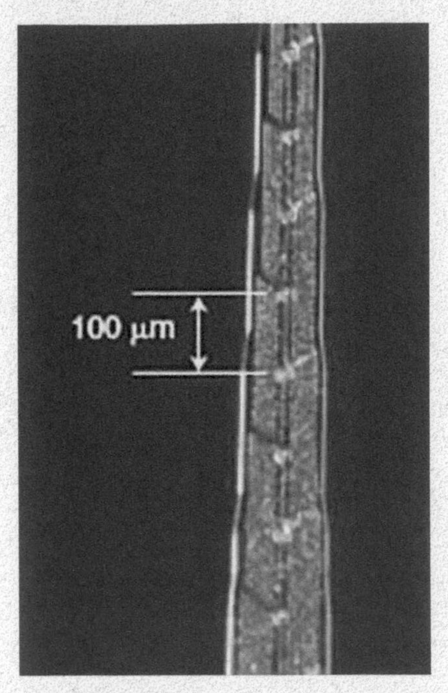

This tiny wire contains devices that help transmit sound. It is attached to the auditory nerve inside the ear.

However, until the scientists in America have finished trials of the new device, the most common treatment for deafness is the hearing aid. There are many different types of hearing aid but most of them work in the same way. They have three main parts.

- A small microphone — this picks up sound waves coming into the ear and converts them into an electrical signal.
- An amplifier — this increases the strength of the signal.
- A small speaker — this transmits the amplified sound into the ear.

Each hearing aid is 'fine tuned' to electronically amplify the range of frequencies most needed by the wearer.

Improved technology has meant that modern hearing aids are small and stylish — they can be mistaken for wireless headsets.

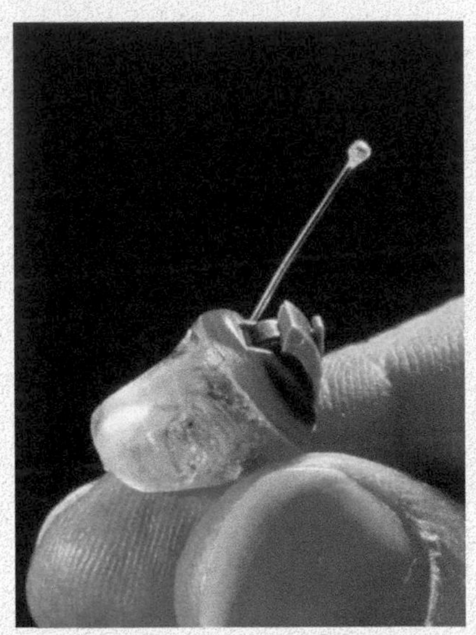

The modern hearing aid fits neatly into the ear canal.

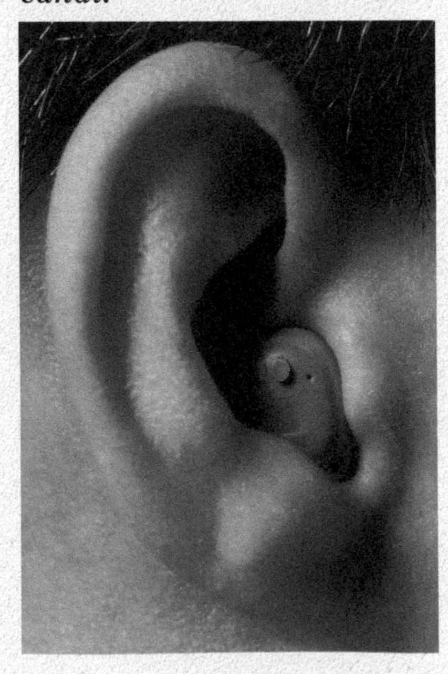

Assess Yourself

1. Why are hearing aids used?
2. Early hearing aids were attached to large batteries that the wearer had to carry with them. They allowed sounds to be heard better than the old-fashioned 'ear trumpets' but they had disadvantages. Suggest what these were.
3. What improvements in technology have resulted in smaller hearing aids?
4. What are the advantages of a smaller hearing aid for the wearer?
5. Describe how the main parts of electronic hearing aids work to help deaf people.
6. What does it mean when we say that a hearing aid is 'fine tuned' for a person?
7. What do you think 'profoundly deaf' means?
8. Describe the treatment that the American researchers are trialling. Does it have any disadvantages?
9. What is the advantage of the American device?
10. What sort of technical difficulties do you think will need to be overcome if the American device is to be used in people?

Citizenship Activity

- Many people suffer from deafness. What sorts of treatment are currently available to them? Consider some of the advantages and disadvantages of each of these types of treatments.
- How could you find out how many people are affected. (Remember that this will not just be the people who are profoundly deaf.)
- In your local area find out how people can begin to learn sign language. How could you improve the availability of any training?

Prepare a brief report on what you have found out.

Level Booster

8 Your answers demonstrate an extensive understanding of sound, its application in a social context and the development of technology to utilise this.

7 Your answers show an advanced understanding of sound and an advanced appreciation of how scientific progress helps to solve social problems.

6 Your answers show a good understanding of sound and hearing, a good grasp of scientific terminology, and a good appreciation of how science is used in technology applications and its effect on peoples' lives.

5 Your answers show a good understanding of sound and hearing, a good grasp of some scientific terminology, and some appreciation of how ideas are applied in a practical context.

4 Your answers show a basic understanding of hearing and a basic grasp of some scientific terminology.

Heat and temperature

BIG IDEAS

You are learning to:
- Recognise temperatures of some common objects
- Understand the difference between heat and temperature
- Understand how temperature is measured 

4

Hot and cold

When we describe an object as **hot** or **cold** we are comparing it with another object. For example a warm room is hot compared to the inside of a fridge but it is cold compared to a sauna.

When we say we feel hot, or that a cup of tea is hot, or that the surface of the Sun is hot we mean different things each time.

Heat is a type of **energy**.

Temperature

Temperature is a way of measuring how hot or cold something is against an agreed **scale**. This common scale allows us to compare the temperature of different things.

Temperature is measured in the units **degrees Celsius** using a **thermometer**. A thermometer is usually made from glass and measures temperatures between 0 °C (the temperature at which pure water freezes) and 100 °C (the temperature at which pure water boils).

1. What is the unit of temperature?
2. Why is it useful to have a common temperature scale that has fixed points on it?

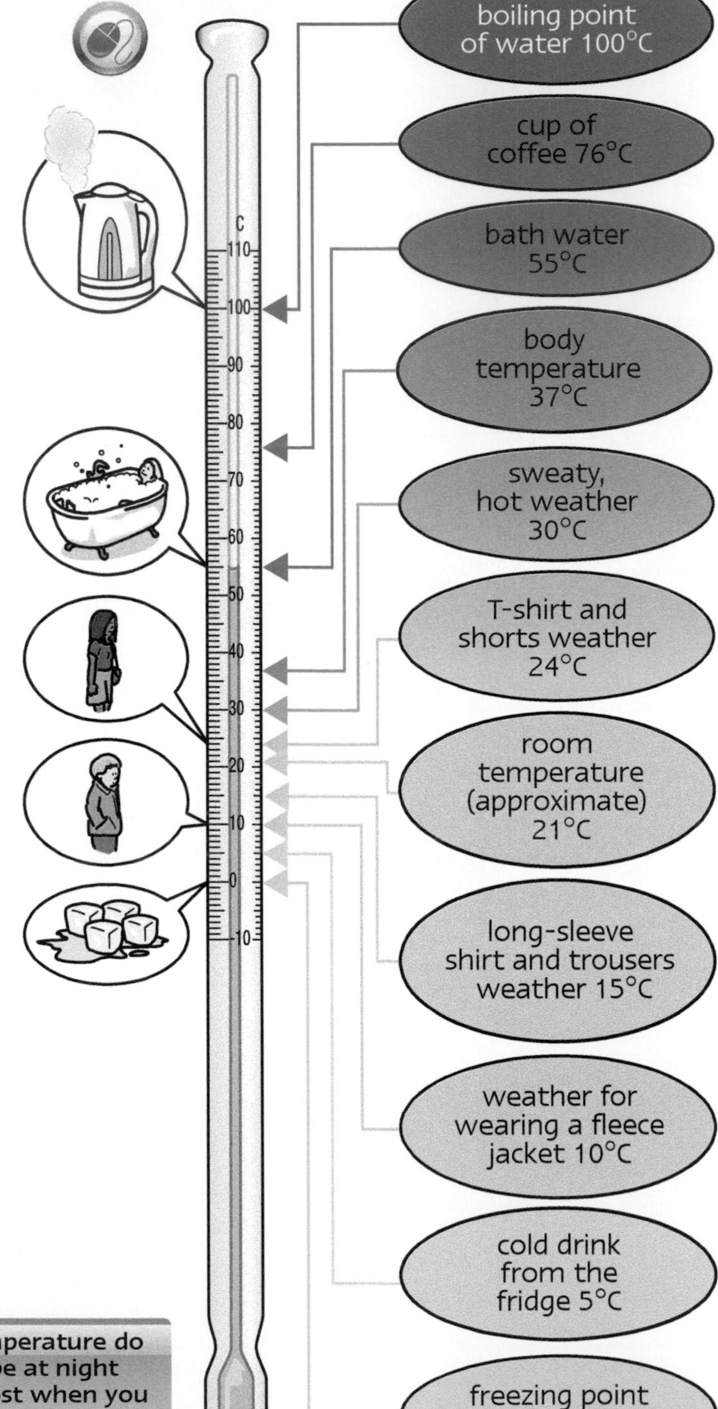

FIGURE 1: What temperature do you think it has to be at night for there to be a frost when you wake up in the morning?

... absolute zero ... cold ... degrees Celsius ... energy ... flow

Different types of thermometers such as clinical thermometers, cooking thermometers and gas thermometers are used to measure very hot or very cold temperatures. Temperature is a measure of how hot something is. Heat is a measure of how much energy an object has. There are many events that happen below 0 °C and above 100 °C. The coldest temperature possible is called **absolute zero** and is −273 °C. Nothing can be colder than absolute zero.

3 Why are different types of thermometers used to measure more extreme temperatures? HSW

4 You are asked to design a thermometer capable of measuring very cold temperatures. What would be its main design features? HSW

How Science Works

Scientists often have to measure temperatures in very cold places. Use the Internet to research how they do this. HSW

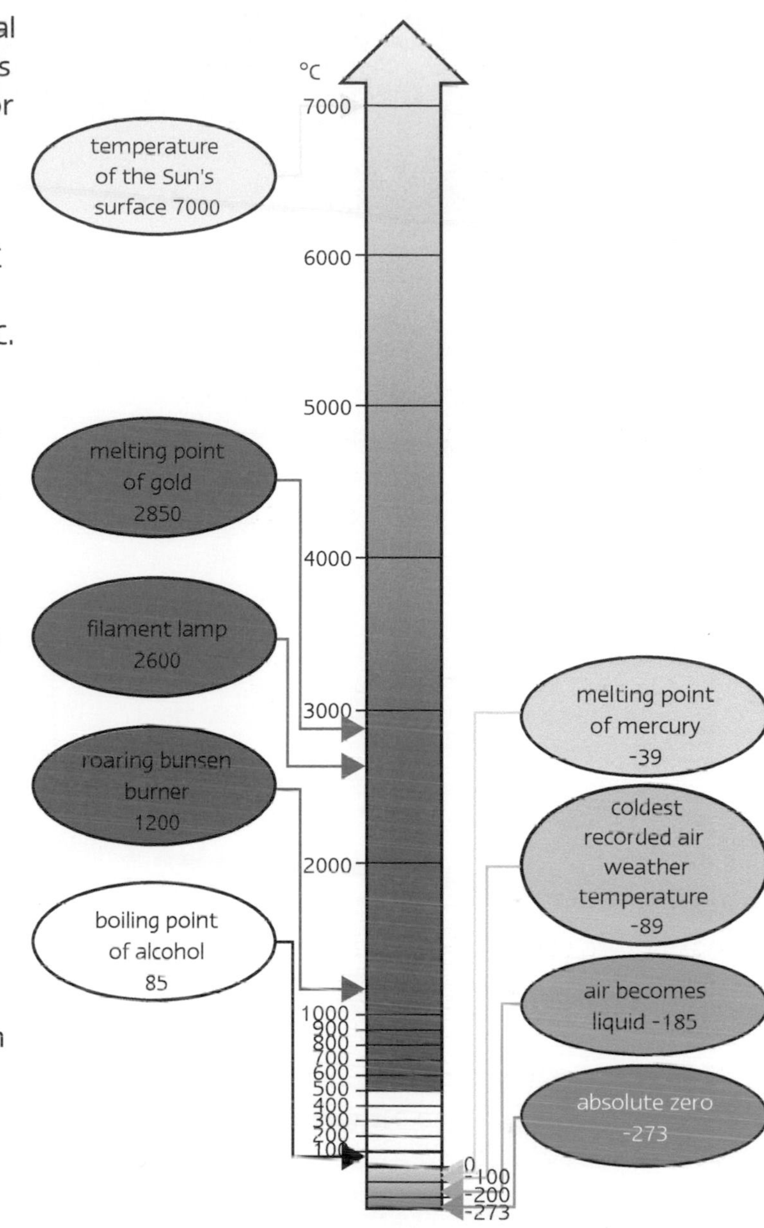

FIGURE 2: What is the melting point of alcohol? (*Hint:* the melting point and freezing point of a substance are the same.)

How heat flows

If there is a difference in temperature between an object and its surroundings or another object then there is a **flow** of heat. Heat always flows from where it is hotter to where it is cooler. If an object is placed in warmer surroundings heat flows into the object from the surroundings. If an object is placed in cooler surroundings heat flows from the object into the surroundings.

5 Why does bath water get colder?

6 A baked potato is taken out of the oven and left on a table. Its temperature is 80 °C.

a What is the temperature of the potato the next morning?

b Explain what happens to the heat from a hot potato left to cool.

7 Explain why a candle has a higher temperature but less heat energy than a bowl of warm washing-up water.

Did You Know...?

- The coldest recorded air temperature in the world is −89 °C in Vostok, Antarctica, in July 1983.
- The hottest recorded air temperature in the world is 58 °C in Al-Aziziyah, Libya, in September 1922.

Getting warmer

BIG IDEAS

You are learning to:

- Identify different ways of heating things
- How heaters make things hotter
- Explain how a microwave oven works

Using heat from the Sun

If you lie on a beach in summer you soon get hot. The Sun's energy gives us both heat and light. Unfortunately the Sun cannot directly provide all the heat we need all the time.

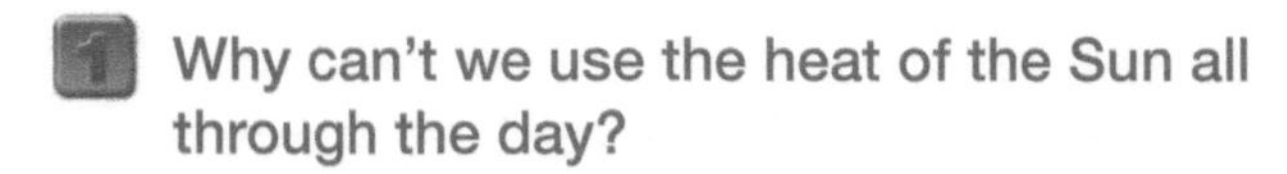

1. Why can't we use the heat of the Sun all through the day?

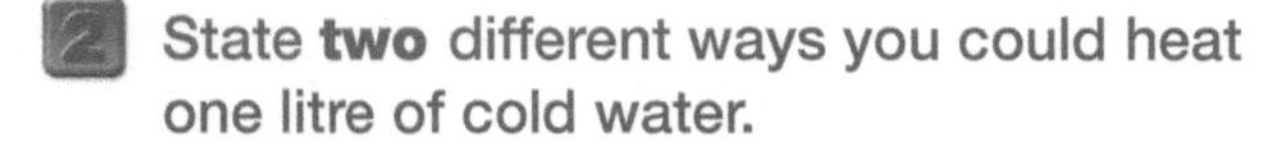

2. State **two** different ways you could heat one litre of cold water.

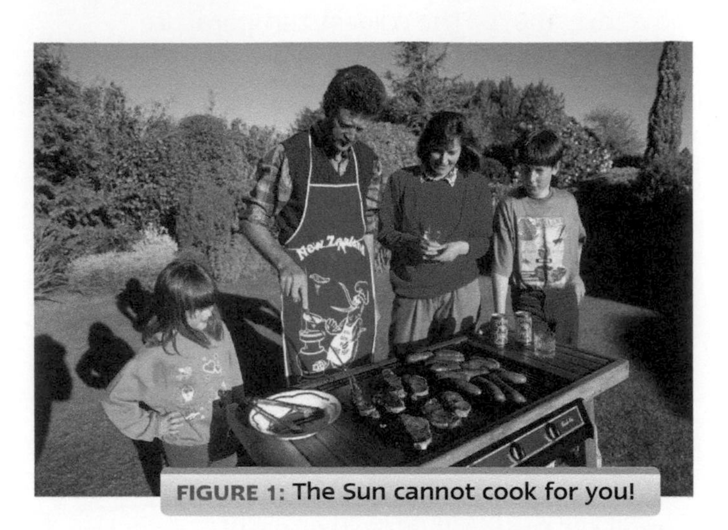

FIGURE 1: The Sun cannot cook for you!

Using infra-red as heat

If you touch an ordinary light bulb when it is on, you can burn yourself. Most of the electricity is transferred into heat energy. Only a small amount of the energy goes to produce light energy.

Some lamps are specially designed as heaters. We call them **infra-red** lamps.

3. Why is it safer to evaporate alcohol using an infra-red heater rather than using a Bunsen burner?
4. Why do athletes use infra-red lamps?

How Science Works

Some athletes use infra-red lamps to soothe aching muscles after training sessions. Infra-red lamps are also used to keep new-born farm animals warm.

Keeping heat inside

Most houses have a hot water tank. This tank is usually **insulated** to keep the water hot. The water can be heated using an electrical **immersion heater**.

The immersion heater can be switched on and off to give hot water when it is needed. As the electricity flows through the heaters they heat up. The heat energy is used to warm up the surrounding water in the tank (see page 156).

 **5** A person can save money by not buying insulation for their hot water tank.
Do you think this is a good idea? Explain your answer.

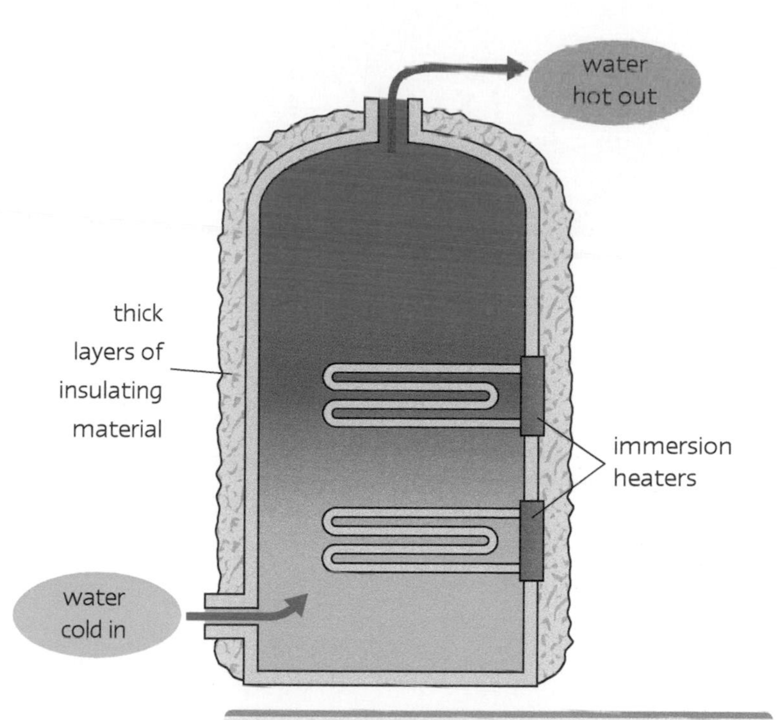

FIGURE 2: Electrical immersion heaters are used in houses. Why is there a thick layer of insulating material around the tank?

Microwave ovens 

Using a **microwave** oven is a quick and easy way to cook food or to heat up a drink or some soup. All of these foods have something in common. They all contain water.

Just as a musician can tune a musical instrument, microwave ovens are tuned to act as heaters. The microwave energy is set just right to be absorbed by the water in food. As the water particles gain energy, they bump into each other and the heat spreads. This is how the food becomes hot and cooks right through.

6 Porridge is made from dry oatmeal and milk. Why couldn't a microwave easily heat up oatmeal on its own?

7 What are the advantages and disadvantages of using microwave ovens to cook food?

Did You Know...?

Microwave ovens are surrounded by a metal grid that stops the microwaves escaping. You can often see the metal grid in the glass door.

FIGURE 3: A microwave oven. Can you explain how cooking with microwaves works?

Conduction

BIG IDEAS

You are learning to:

- Recognise difference between conductors and insulators
- Explain conduction using a particle model
- Describe some uses of conductors and insulators 

Good conductors

Heat can move from one place to another.

When a metal poker is used in a coal fire the heat travels through it very easily and the poker can burn a person's hand. The poker cannot be held in the fire for long. It is a good **conductor** of heat.

When kebabs are cooked on a barbeque they have a metal skewer placed through them. This helps the food to cook more quickly.

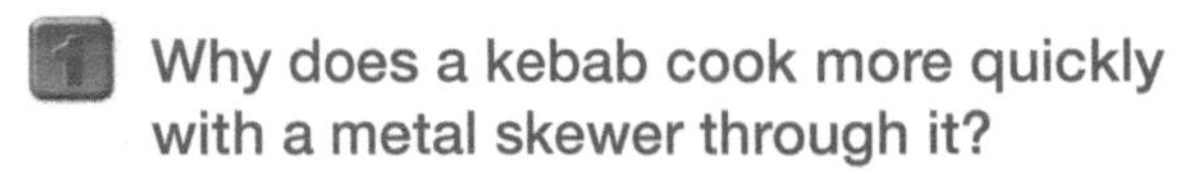

1. Why does a kebab cook more quickly with a metal skewer through it?

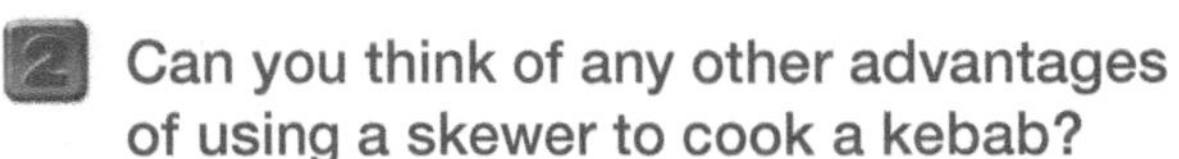

2. Can you think of any other advantages of using a skewer to cook a kebab?

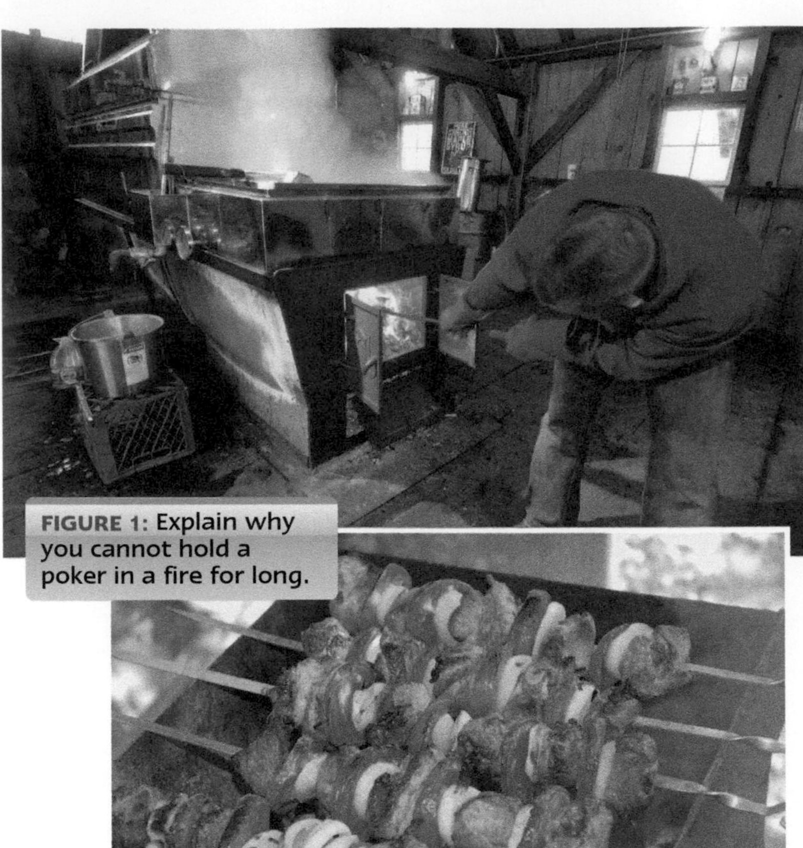

FIGURE 1: Explain why you cannot hold a poker in a fire for long.

Insulators

If heat energy moves through an object or material from where it is hotter to where it is cooler without the object itself moving, then we say that the heat has been conducted through the material. An object that does not allow heat to pass through it easily is called an **insulator**.

3. What is the difference between a conductor and an insulator?
4. Explain what conduction means.
5. List **three** materials that are good insulators.

Did You Know...?

A metal object feels hotter or colder to the touch than a wooden or plastic object at the same temperature. This is because it conducts the heat into or out of your hand very easily and not because it is colder or hotter.

... air ... conductor ... gas ... heat ... insulator

Conduction 

Conduction mostly happens in **solids**. If a solid allows heat to pass through it easily then it is a good conductor of heat. Most metals and stone are good conductors of heat. Materials such as plastic, wood and glass are poor conductors of heat.

Many common objects are carefully designed to be either good conductors or good insulators. A metal saucepan often has a plastic handle. The metal allows the heat from the cooker ring to pass easily into the food and the plastic handle prevents a person's hand from being burnt when they pick up the saucepan.

Liquids and **gases** are poor conductors of heat. Insulators can be used to prevent heat passing through a material.

6 Make a list of common materials that let heat conduct through them easily.

7 Make a list of some common items that are designed to act as insulators.

Conducting particles

Solids are good conductors of heat because the **particles** in a solid cannot move. When one end of a solid rod is heated the particles at that end gain heat and start to vibrate. The vibrating particles bump into their neighbours and pass the energy on. This process continues throughout the material until all the particles are vibrating by the same amount. Gradually the extra heat energy is passed all the way through the solid and this causes a rise in temperature at the other end. A lot of the heat is transferred to the surroundings during this process. **Air** is such a bad conductor that it is often used to help keep things warm.

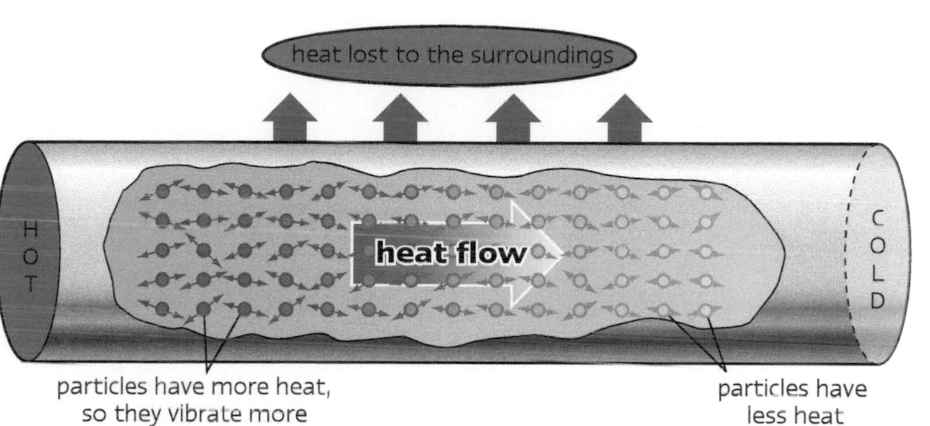

FIGURE 2: Conduction through a solid. Why are solids better heat conductors than gases or liquids?

8 Explain how heat energy is conducted through a solid copper saucepan.

9 Why do you think that gases are poor conductors of heat? (*Hint:* think about how the particles are arranged in a gas.)

10 Explain why the clothing worn by skiers and mountaineers is usually made of feathers and down.

How Science Works

Fire-fighters often wear protective clothing made from an insulating material. The material has lots of tiny air pockets that prevent heat passing through them and help keep the fire-fighter cool. The material used is usually also shiny to **reflect** as much heat as possible.

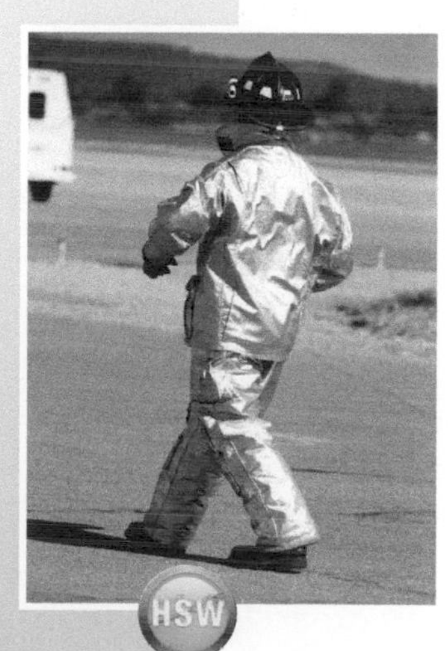

HSW

Convection

BIG IDEAS

You are learning to:
- Describe convection
- Explain convection using a particle model
- Recognise some applications of convection

Convection currents and thermals

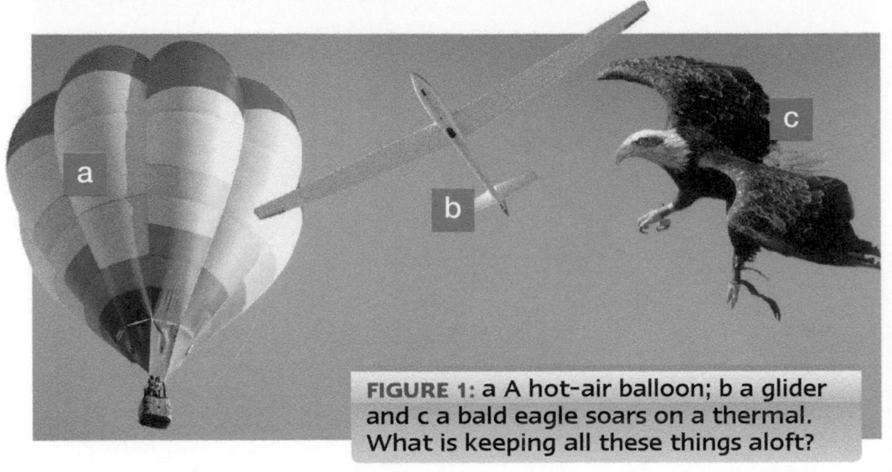

FIGURE 1: a A hot-air balloon; b a glider and c a bald eagle soars on a thermal. What is keeping all these things aloft?

In nature rising hot air creates **thermals**. When hot air rises cooler air falls and takes its place.

When a kettle is switched on the element heats the water that is next to it. The hot water rises and cold water takes its place next to the element. This causes a circulation of water in the kettle until all the water reaches the same temperature. This circulation of water is called a **convection current**.

Convection can only happen in liquids and gases because they can move.

1. What is a convection current?
2. Why is the element in a kettle at the bottom and not at the top?
3. How are thermals useful to birds?

How Science Works

Hang-gliders work by using thermals. The large wing is designed to trap rising air underneath it and make the glider soar higher. They are often launched from hillsides where there is more wind. They will not work on cold days.

Convection and density

Convection is the main way that heat travels through all liquids and gases. It happens whenever one part of a liquid or gas is heated more than the rest. The warmer liquid (or gas) rises as it has expanded when it warmed up. This expansion means that it is less **dense** than its surroundings.

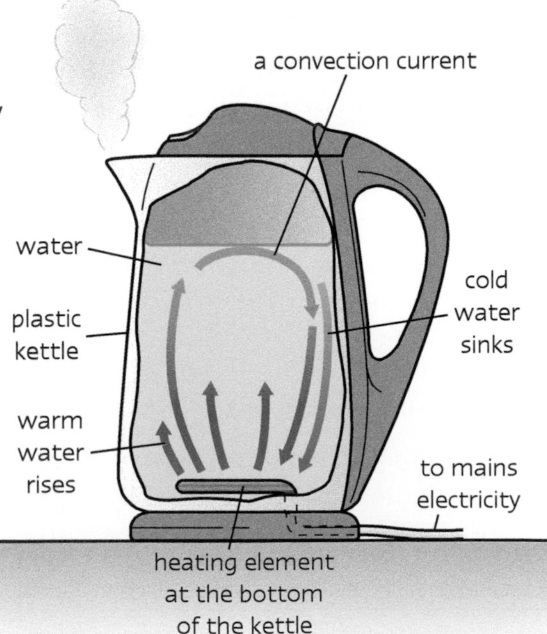

FIGURE 2: A convection current in a kettle. Why is the kettle made from plastic?

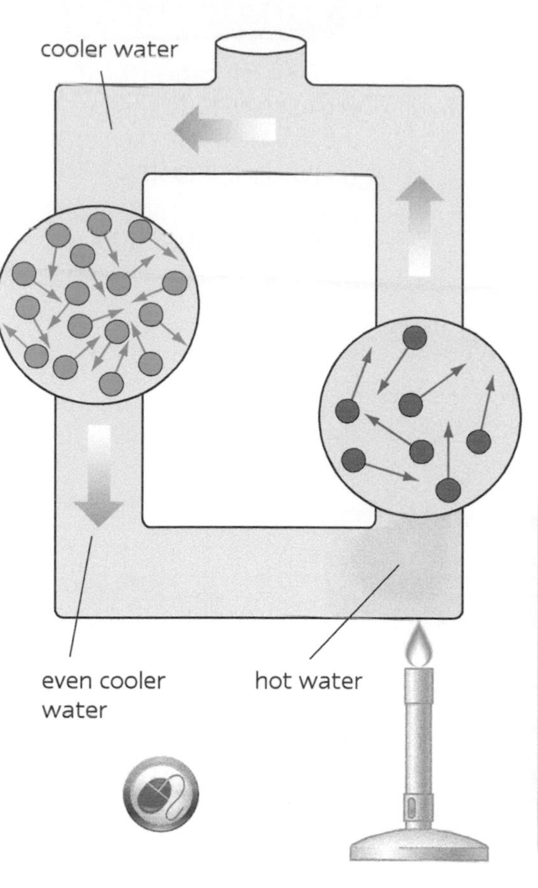

FIGURE 3: In hot water the particles have more energy so they move quicker and are more spread out. Hot water is less dense. When water cools the particles have less energy so they slow down and move closer together. Cold water is denser so it sinks below hot water.

Convection takes place in gases as well as liquids, but it cannot occur in solids. Convection causes wind and ocean currents as the moving gas and liquids carry heat from where it is hotter to where it is less hot.

How Science Works

Modern ovens are fan-assisted. There is a fan at the back of the oven that helps to move hot air away from the element around the oven (red arrows).

These convection currents heat the oven to the same temperature more quickly and save energy.

HSW

The same principle is used in home central heating. A **radiator** or heater warms the air next to it which then rises. Cold air takes its place and eventually the room is heated to a constant temperature.

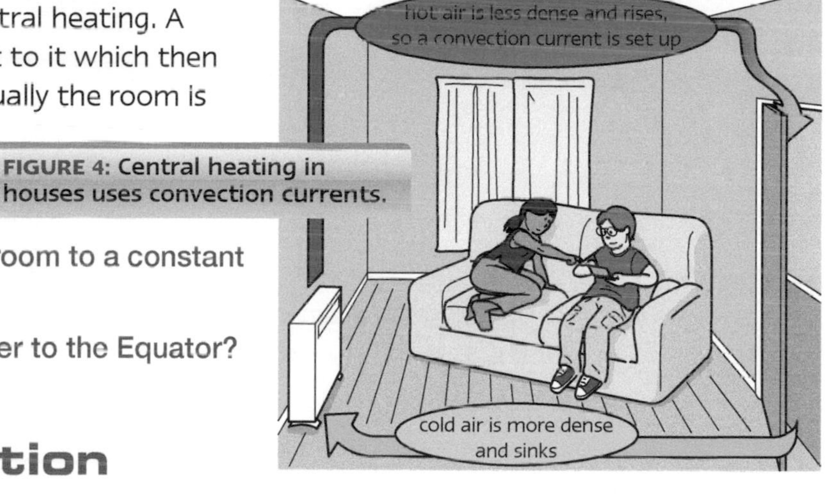

FIGURE 4: Central heating in houses uses convection currents.

4 What does 'density' mean?

5 Describe how a radiator heats a room to a constant temperature.

6 Why are the oceans warmer nearer to the Equator?

Explaining convection

Heat travels from hot areas to colder areas by convection because the **particles** in a liquid or a gas are free to move. Liquids and gases are called **fluids** because they can move. When a fluid is heated, heat energy makes the molecules move faster and the fluid expands. This reduces the density of the warmer part of the fluid. The less dense parts move upwards through the cooler parts. As they move they lose heat to the surroundings and cool down again.

7 Explain why warm air rises.

8 Why does warm air cool as it rises?

9 What might happen if warm air did not cool down as it rises?

10 Why is it very difficult for hot air balloons to attain very high altitudes?

... fluid ... particle ... radiator ... thermal

Radiation

BIG IDEAS

You are learning to:

- Explain how heat energy from the Sun travels to the Earth and what happens to it
- Describe what happens when radiant heat hits a surface
- Describe some uses of radiation or radiant heat
- Explain what radiant heat is

Radiant heat from the Sun

The type of heat energy that comes from the Sun is called **solar radiation** or **radiant heat**. Only a small amount of it reaches the Earth. Cloudy days are cooler than clear days because this type of radiant heat energy travels in straight lines. It cannot go around the clouds. In fact the clouds reflect a lot of the heat back into the atmosphere.

1. Explain why we get shadows on a sunny day.
2. What happens to heat energy from the Sun that does not reach the Earth?
3. Explain why clear nights are cooler than cloudy nights.

FIGURE 1: White and light-coloured clothes reflect heat better.

Reflecting and absorbing radiation

When radiant heat reaches the Earth it is absorbed by the surfaces that it hits making them warm up. The more heat radiation it absorbs the hotter an object gets.

Differently coloured surfaces absorb different amounts of heat.

- White and shiny surfaces reflect more heat than dark or dull surfaces. They are not good absorbers of radiant heat. They do not radiate much heat.
- Dark surfaces are better at absorbing heat and as they are also poor reflectors of radiant heat they heat up more quickly.

All hot objects give off (**emit**) radiant heat. The hotter an object is the more heat it radiates. Light shiny surfaces are not as good as emitting radiant heat as dark or dull surfaces.

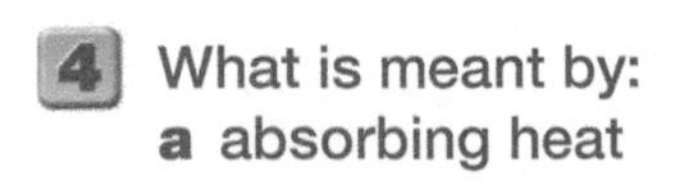

4. What is meant by:
   a absorbing heat **b** reflecting heat?

5 What happens to an object when it absorbs heat?

6 Explain why people living in desert areas usually wear white clothing.

Electromagnetic waves and radiant heat

In space there are no particles present so radiant heat must rely on a different method of energy transfer. **Electromagnetic waves** carry the radiant heat. The waves transfer energy from one place to another without the material it is travelling through moving. Heat radiation is carried by **infra-red** rays and **ultra-violet** rays.

7 What carries heat radiation from one place to another?

8 Why is the inside of an oven painted black and not white?

Did You Know...?

In some hot countries most of the buildings are painted white. The colour white reflects radiation very well and this keeps buildings cooler.
In cold countries buildings are often darker in colour so that they absorb as much heat as possible.

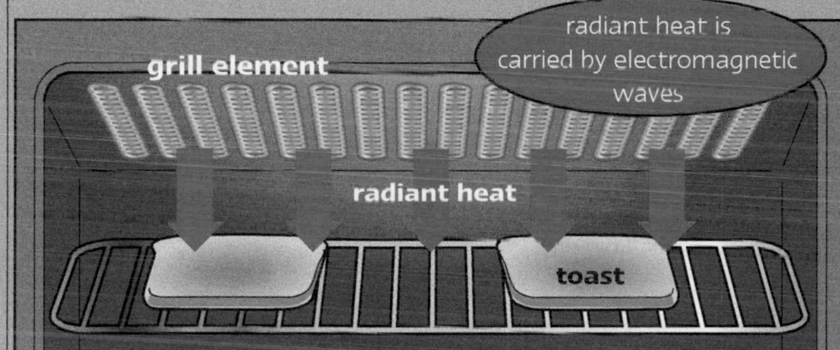

FIGURE 2: When bread is toasted under a grill the heat must travel downwards from the grill elements on to the bread. The bread absorbs this heat energy and warms up and browns. The radiation still transfers the heat energy even though there are air particles present around it.

9 If a leaf falls onto fresh snow, after a few days of sunshine it will have sunk into the snow. Explain how this has happened.

How Science Works

Infra-red detectors are used by the emergency services to find people who are trapped, especially after an earthquake. The heat from their bodies is detected by a special camera. The camera produces a heat picture of an object called a **thermogram**.

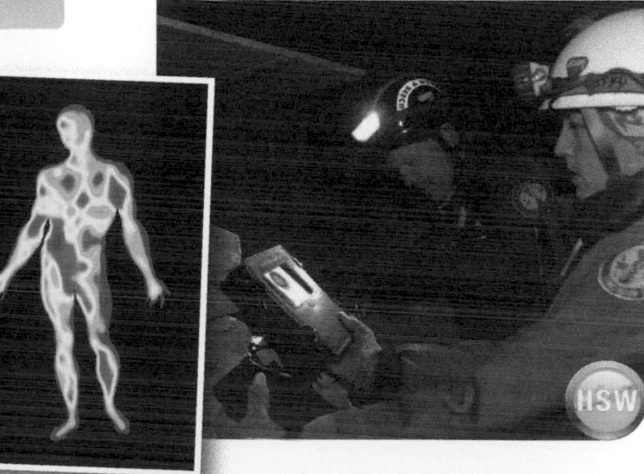

... solar radiation ... thermogram ... ultra-violet

Practice questions

1 Complete the sentences in your exercise book. Use words from the list.

cochlea **eardrum** **ears** **particles** **middle** **nerve**
signal **vacuum** **vibrate** **vibrating** **vibrations** **wave**

All sounds are made by _________ objects. Sound travels as a _________ from a source to our _________. The sound can travel through solids, liquids and gases but not through a _________. A vacuum is where there are no _________ present. When they reach our ears they can make the _________ vibrate. The _________ of the eardrum are then passed onto the three small bones in the _________ ear. From there the vibrations are detected by the _________ causing it to _________. The auditory _________ detects the vibrations and sends a _________ to the brain.

2 Noise pollution is used to describe any unwanted sound.

- **a** Give **three** examples of noise pollution.
- **b** Describe **one** way of reducing noise pollution.
- **c** Our hearing can be damaged by loud sounds. What two things can we do to prevent this happening.

3 Which is the most accurate statement?

- **a** Heat and temperature are the same thing.
- **b** Heat is how hot something is.
- **c** Temperature is measured in joules.
- **d** Temperature measures how hot an object is.

4 **a** What is meant by conduction?

- **b** Give **two** examples of when conduction can be useful
- **c** Convection is another way that heat can travel. What is meant by convection?
- **d** Why does convection not happen in solids?

5 Heat transfer methods that involve particles are

a Conduction and radiation.

b Conduction, convection and radiation.

c Conduction and convection.

d Convection and radiation.

6 Sound travels fastest in solids because the particles are closely packed and cannot move. Explain, using ideas about particles why sound waves travel faster in liquids than they do in gases.

7 Explain the following statements

a Why is it usually warmer during the day when the sky is clear and not cloudy.

b Why does a cold drink from the fridge quickly warm up if it is left in the sun.

c Athletes often wrap silver blankets around them after a race.

d Fish and chips are often wrapped in newspaper or put in cupboard boxes.

8 Explain why sounds become fainter as you move further away from the source.

9 Which two statements are correct?

a High pitch sounds have low frequencies.

b Low pitch sounds have low energy.

c Frequency and pitch are different things.

d High pitched sounds have high frequencies.

e Low pitch sounds have low frequencies.

f High pitch sounds have low energy.

10 Explain using ideas about energy transfer, why a beaker of ice left in a room will melt and warm up, but will not get any hotter than the room it is in.

11 Explain using ideas about energy how a double glazed window reduces heat transfer. Why will some energy always get through?

Learning Checklist

4

- ☆ I know that sounds are made by vibrating objects. page 132
- ☆ I know that sound becomes fainter further away from the source because energy dissipates or spreads out. page 136
- ☆ I know that sounds travel at different speeds in solids, liquids and gases. page 140
- ☆ I can give some examples of common temperatures on the Celsius scale. page 150

5

- ☆ I know how to calculate the speed of sound. page 137
- ☆ I know what an echo is and I can explain what causes it. page 137
- ☆ I can describe how insulators can reduce heat loss. page 153
- ☆ I can describe conduction, convection and radiation. pages 154–159

6

- ☆ I know the difference between a compression and a rarefaction. page 138
- ☆ I can give some uses of an ultrasound. page 143
- ☆ I can use the particle model to explain conduction and convection. pages 155, 157
- ☆ I can describe how temperature difference leads to a flow of energy. page 157

7

- ☆ I can use a model ear to describe some possible causes of hearing impairment. pages 144–146
- ☆ I can use my ideas about heat transfer to explain the use of conductors and insulators in a range of situations. pages 153–154
- ☆ I can use my ideas about energy to explain heat transfer by radiation. page 158

8

- ☆ I can explain using ideas about energy transfer, the dissipation of heat and sound during transfer processes. pages 136–159

Topic Quiz

1. What is the common unit of temperature?
2. What is room temperature?
3. What is the boiling point of water?
4. What is the difference between heat and temperature?
5. What is an insulator?
6. Why is the inside of an oven painted black?
7. What causes a sound?
8. How can a guitar produce a higher note?
9. What is the range of human hearing?
10. Why can sound not travel through a vacuum?
11. What are the units of frequency?
12. Why do sounds become quiet further away from their source?
13. What is ultrasound? Give an example of its use.

True or False?

If a statement is false then rewrite it so it is correct.

1. Heat travels in solids by convection.
2. Energy from the Sun reaches us by conduction.
3. Heat energy flows from hotter places to cooler places.
4. Dark surfaces are good at reflecting heat energy.
5. Double glazing works by preventing conduction and radiation.
6. Sound travels fastest in liquids.
7. Sound can travel through a vacuum.
8. Pitch and frequency are the same thing.
9. Ultrsound is used for cooking.
10. A compression is a low pressure where the particles are spread out.
11. Loud sounds have a larger amplitude than quiet sounds.
12. The ear drum is the only part of the ear that vibrates.

Literacy Activity

Write an article about saving heat energy at home.

You should explain what methods you could use, why they work and why you need to do it.

You could choose to do this as

- An article in a local newspaper
- As a poem
- As a poster for display in your school
- For pupils in a primary school
- As a leaflet for local people

Write a letter to a newspaper about noise pollution. You need to explain some of the concerns that people have about their exposure to noise pollution. You should also consider some possible solutions and decide if these are useful or not.

ICT Activity

Make a short powerpoint presentation to teach year 5 and 6 pupils some of the key facts about sound that they should know.

Who killed cock robin?

BIG IDEAS

By the end of this unit you'll be able to explain how energy is transferred in food webs, how the resources in a habitat affect the numbers and how habitats change. You'll be able to explain the applications and implications of these ideas.

A cock robin has been found dead – but how did it die? It is 15 times more likely that the robin was killed by a domestic cat than a sparrow hawk or owl. If the robin has shed feathers, it might have died from a fungal infection. It is also possible that it was a road casualty or died defending its territory from another male robin.

The male robin is a very aggressive animal and has been found to swoop on and viciously peck red dummy birds left in its territory. Robin numbers are in decline and in very cold winters they are likely to die from the cold or starvation if they cannot get enough food. On a cold night a robin can lose up to 10% of its body weight and will need to replenish its reserves within two days. Putting food out on a bird table in the winter can help the robin survive.

So what else do we know about the robin?

Breeding

The male robin starts courtship at the end of January and breeds in March when he has established a territory. The territory is about 0.5 hectares (six territories would fit into a football pitch). A robin nest is cup-shaped and is made from dead leaves, grass, hair and mosses. It can be found in hedges, walls, holes in trees and even garden sheds. The robin lays about four to six eggs in a clutch and they are incubated over the next 13 days. Usually two clutches are produced in a year, although in good conditions four can be produced. The robin starts to reproduce after one year.

Age

The record age for a robin is 17 years although the normal life expectancy is two years.

Feeding

Robins feed on seeds, fruit, worms and insects and will take fat, cheese, biscuits, peanuts and mealworm from a bird table. Robins are one of the more common garden birds and with a little patience they can be fed by hand.

Song

Robins sing throughout the year, mainly during the courtship period. Both the male and the female sing during the evening and even in the middle of the night.

What do you know?

1 Construct a simple food web to show what the robin feeds on and what feeds on it.

2 What evidence is there that the robin is an omnivore?

3 The robin is a bird. How is it adapted to fly?

4 Which factors affect the population size of robins?

5 How can a person help to support the robin population?

6 Why are robins more likely to die from starvation in the winter?

7 Why does a robin have a red breast?

8 If all the robin eggs from a clutch survived to produce adults, how many adults would you expect to be produced in a year?

9 Why do so few robins survive?

10 The size of a robin's territory varies according to how much food is available. Why would the territory be larger in very dry seasons?

Design a predator

BIG IDEAS

You are learning to:
- Identify key features for survival
- Explain why the key features are important to an animal's survival
- Explain the benefit of **protective colouration**

5

Features for survival

The table gives some of the common features for **survival** shown by predators and prey.

Eyes at the front of the head to judge distance	Eyes at the side of the head to give all round vision
Strong muscles to give rapid acceleration	Strong muscles to allow the animal to run fast for a long time
Excellent sense of sight	Excellent sense of hearing
Sharp teeth	Sharp and strong claws/talons
Able to keep still for a long period of time	Well camouflaged
Excellent sense of smell	Produce poisons

1 Study the list and then for each of the animals listed, pick out **three** key features that are important for its survival. Explain why you have chosen each feature.

a Tiger
b Pike (live in freshwater ponds and are the top carnivore)
c Eagle
d Zebra (eaten by lions)
e Cobra

6

Protective colouration and mimics

Wasps and bees are brightly coloured with yellow and black stripes. They feed on nectar and pollen from flowers. Both animals have stings. The bee usually can only sting once. The drone fly also has yellow and black stripes and it feeds from flowers. It has no sting. Young toads feed off flies and will try a wasp or bee once but never again.

2 Explain why the wasps and bees are brightly coloured and how this helps the survival of the species.

3 Why is a drone fly brightly coloured even though it cannot sting?

... mimic ... protective colouration

Shake and rattle

Many snakes are poisonous but very few of them are brightly coloured. The rattle snake lives in desert areas. It is a sandy colour to match its background. It feeds on small mammals. When it is disturbed it will shake the rattle on its tail rapidly to make a sound.

4 Explain why the rattle snake is camouflaged and how the rattle on its tail aids its survival.

5 Some poisons stop an animal's muscles working. Why does the animal die?

Did You Know...?

Some of the most poisonous animals in the world are found in Australia. The blue-ringed octopus, the funnel web spider and the box jellyfish can all be deadly.
The Indian Taipan is the world's most poisonous snake; one bite contains 110 mg of toxin – enough to kill up to 100 people.

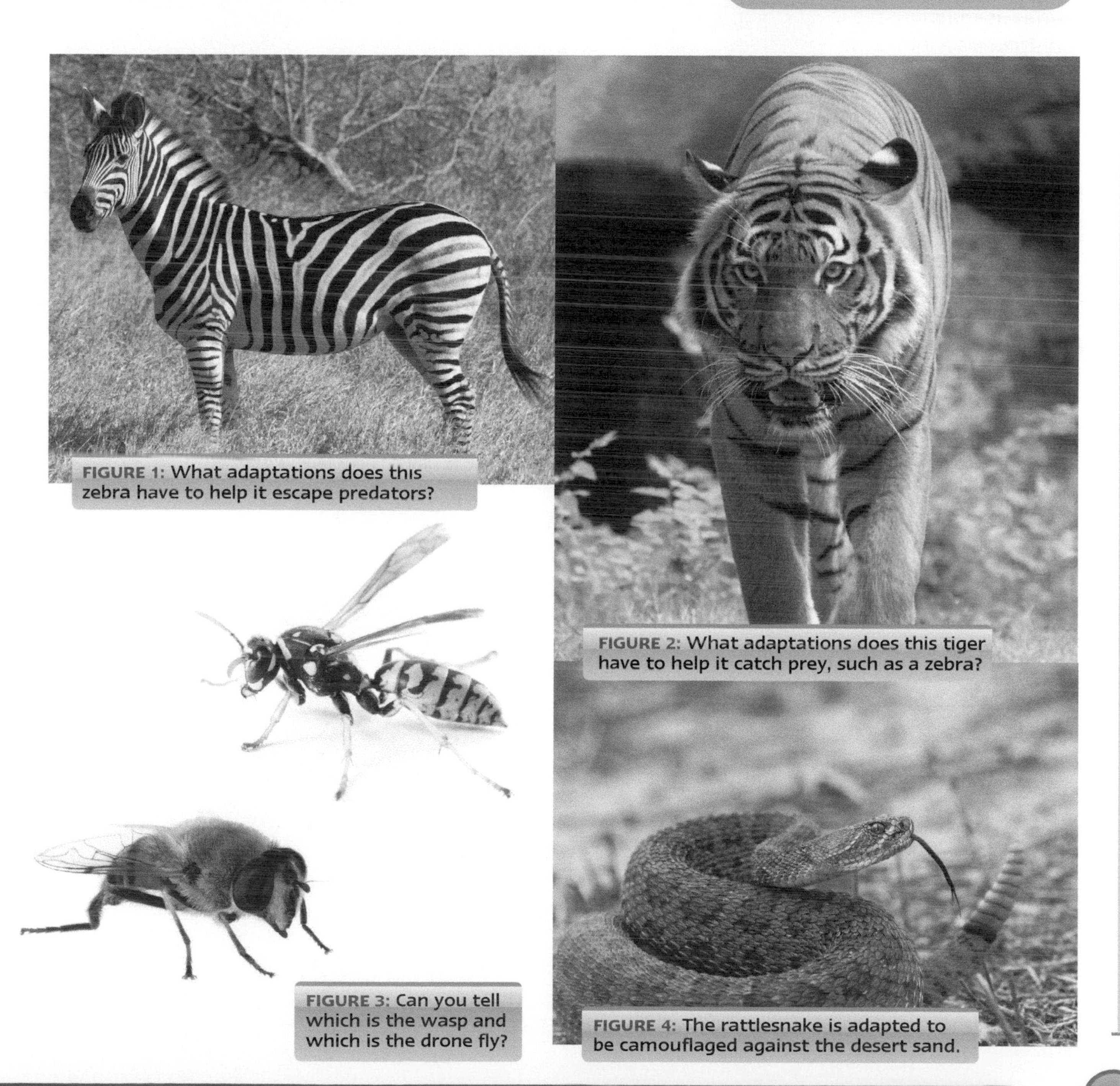

FIGURE 1: What adaptations does this zebra have to help it escape predators?

FIGURE 2: What adaptations does this tiger have to help it catch prey, such as a zebra?

FIGURE 3: Can you tell which is the wasp and which is the drone fly?

FIGURE 4: The rattlesnake is adapted to be camouflaged against the desert sand.

Where has the ox gone?

BIG IDEAS

You are learning to:

- Interpret food webs in terms of energy flow
- Construct models to represent pyramids of numbers 
- Explain why there are fewer top carnivores than other organisms

Energy transfer

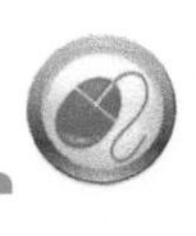

Just looking at the example of one animal can show us a lot about energy flow. Look at Figure 1. Energy is transferred from the sun to plant material by the process of photosynthesis. The grass grows; is eaten by the ox and some of the energy in it is transferred to its body as new tissue, or to create heat or support movement. Not all of the energy is extracted; some remains in the dung as chemicals.

FIGURE 1: Where does this ox get energy from?

The ox is killed by a lion. The lion does not eat the entire ox; some of the meat and internal organs, bones and hooves are left. Hyenas and vultures will wait until the lion has finished and then eat what is left. These animals are called **scavengers**. The hyena has strong, powerful back teeth which can easily crack bones and it has strong acid in its stomach which helps digest bones, skin and even hooves.

Dung beetles roll the **dung** into balls and move it away to bury it under the ground as a source of food for their offspring. They are decomposers.

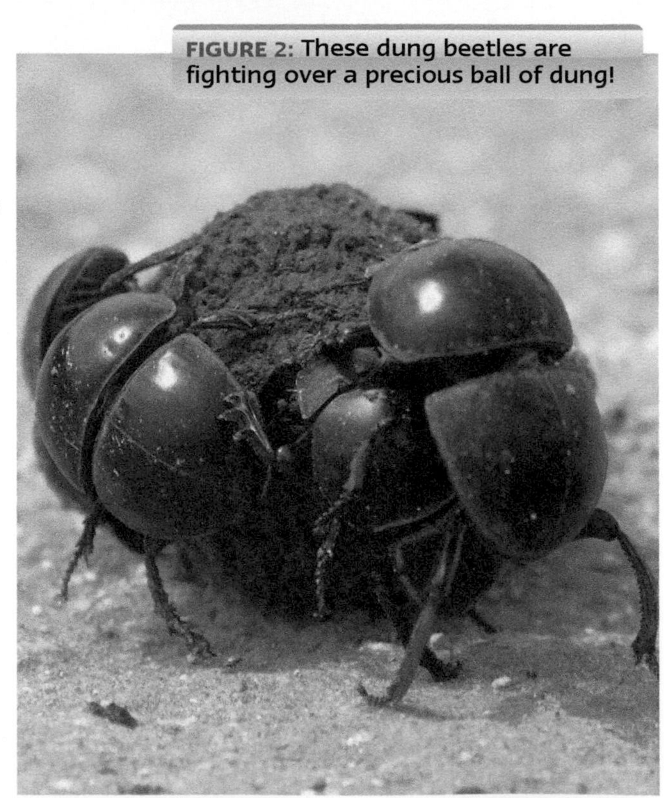

FIGURE 2: These dung beetles are fighting over a precious ball of dung!

1. Construct a food web to show the animals feeding on the ox and its faeces.
2. How is the hyena adapted to eat bones and hooves?

How much energy is lost?

At each stage of the food chain not all the energy is transferred. In fact the carnivore might only receive 10% of the energy in the prey. The rest of the energy does not disappear – it is transferred to different food chains involving scavengers and decomposers. When an organism respires, some of the energy is transferred to the surroundings as heat. This can be lost from the food web.

> **Did You Know...?**
>
> Dung beetles are eaten by burrowing owls. The owls have learnt to put pieces of dung out to attract the beetles.

3 Explain why not all the chemical energy is passed from a prey to a predator.

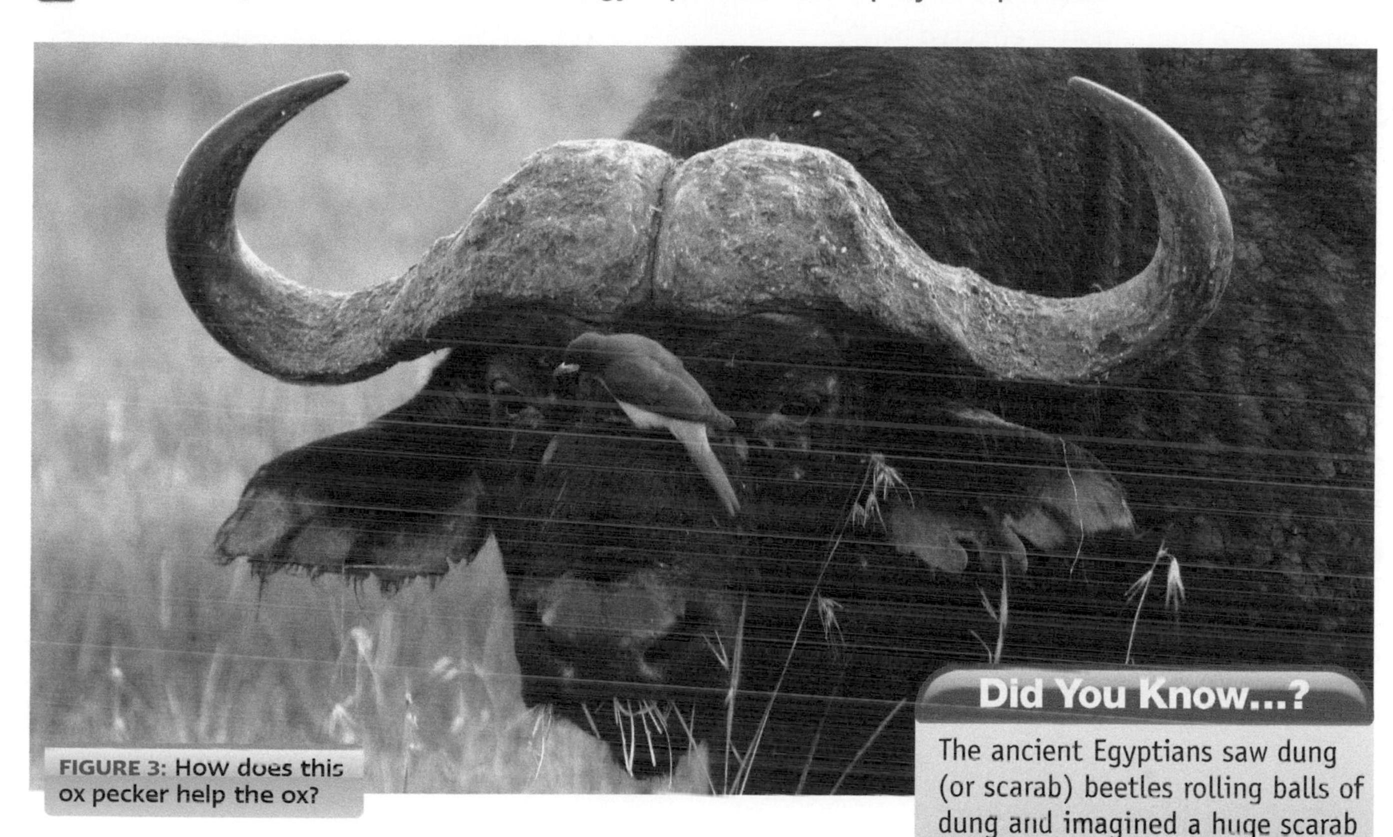

FIGURE 3: How does this ox pecker help the ox?

> **Did You Know...?**
>
> The ancient Egyptians saw dung (or scarab) beetles rolling balls of dung and imagined a huge scarab beetle rolling the Sun across the sky. They associated the scarab with the Sun god Ra and scarabs can be seen in many ancient Egyptian carvings

The ox pecker

An ox pecker is a bird that eats ticks and blotflies that have dug into the ox's skin to feed off its blood. It has a very sharp bill to cut open the skin to reach the parasites. The ox pecker also warns the ox of the approach of predators.

4 Construct a pyramid of numbers for the food chain:
Grass ⟶ Ox ⟶ Ticks ⟶ Ox pecker

5 Explain why there are few top carnivores like lions.

6 Use the model of energy flow through the food chain to explain what a pyramid of energy might look like for the food chain: **Grass ⟶ Ox ⟶ Ticks ⟶ Ox pecker**
Give reasons for your answer.

> **How Science Works**
>
> More food (chemical energy) can be produced from the same area of land by growing plants and eating them, than by keeping animals and eating them. Use the model of energy flow to explain why. Should we all become vegetarian? HSW

Population models

BIG IDEAS

You are learning to:

- Identify the key factors that affect the size of a population
- Use models to describe changes in population (HSW)
- Predict the effect of key factors on the growth of a population

Population size

The size of a population is affected by:

Physical factors
Temperature
Rainfall
Acidity

Catastrophes
Flood
Fire
Drought
Earthquake

Biological factors
Competition
Predation
Disease/Parasites

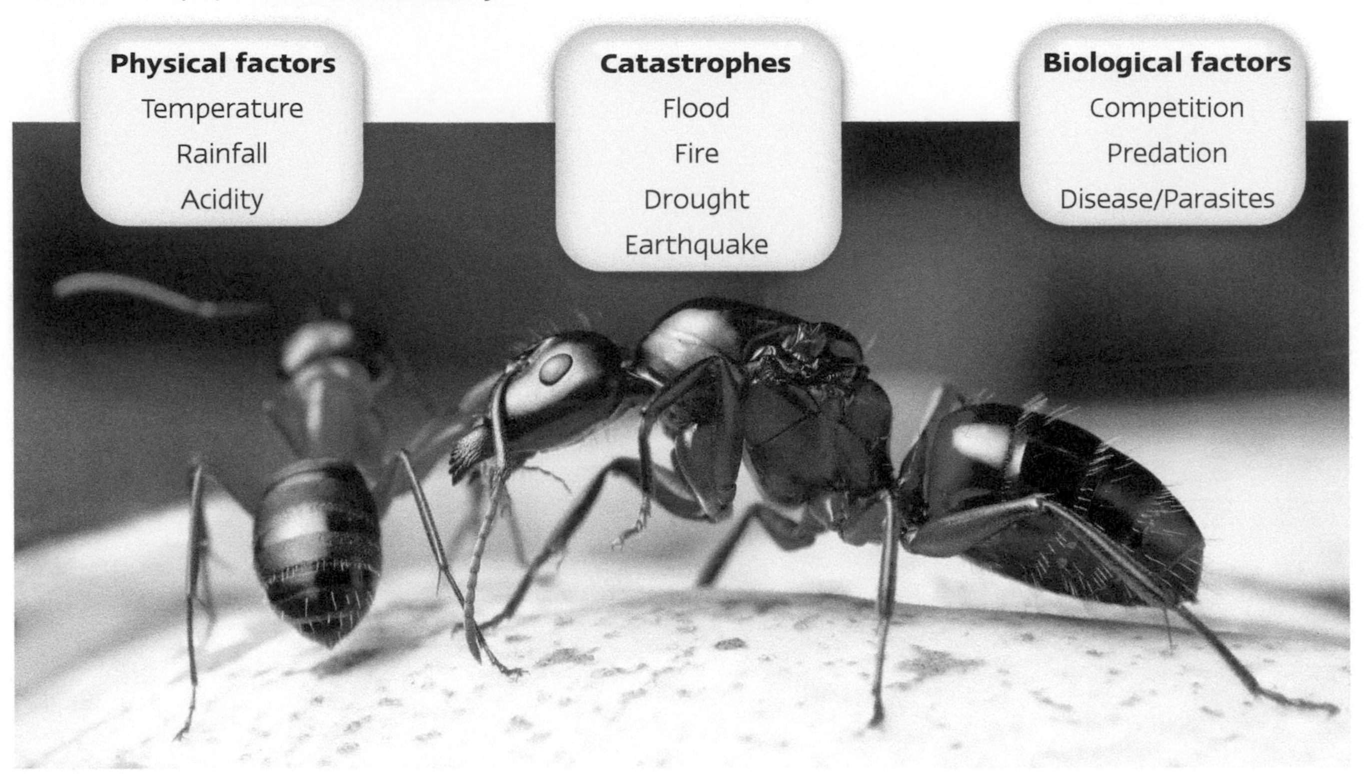

1. Name **two** things that organisms compete for.
2. Give **one** other major catastrophe that affects population size. 

The Year of the Ant

Computers are often used to model population growth. A simple program will work on the idea that growth rate can be worked out using the following formula:

growth rate = birth rate − death rate

Dr Spotswood fed in information for the computer to work out how the **population growth rate** of an ant population changed through the year. He found out that the queen ant is the only ant that produces eggs and the worker ants look after the queen and the colony.

FIGURE 1: Which ant produced these eggs?

a In spring the queen ant lays lots of eggs as she emerges from dormancy in the winter. Several of the eggs hatch into worker ants after 20 days.

b Plenty of food is available. The worker ants feed the queen so more eggs are laid. More worker ants hatch.

c In summer, drought conditions occur and little food is available for the queen to produce eggs. Several workers die.

d The heath land catches fire, killing many workers but not the queen.

e Rain allows new growth of plants and plenty of food is available to feed the queen so she can produce eggs.

f The onset of cold in late autumn kills many workers and the queen prepares to survive the winter in a dormant state.

3 **For each statement a–f, indicate whether you think the population will grow, stay the same or reduce.**

4 **Sketch a simple graph of the changes in population size throughout the year. Label the graph to show where each change occurs.**

Complex models

The use of a simple model can only indicate a rough idea of what might happen. Many other things must be considered to make a more accurate prediction. These include:

- sex ratio of males and females
- number of animals at each age (**age structure**)
- **migration** and immigration.

5 **What is the difference between migration and immigration? How will they affect the population size?**

6 **Why do you not need to consider the sex ratio for an ant population?**

Managing fisheries

The amount of fish such as cod that can be taken from the sea is worked out using computer models. If too many fish are caught, the fish population will fall and in a few years not many fish will be left to catch. Fishermen earn a living from fishing and if they are not allowed to catch enough fish they will not be able to carry on fishing. Modelling is therefore very important in working out catch size.

7 **Study the graphs of two fish populations for a fishery. The fish chosen reproduces at four years of age. Are they being overfished? Give reasons for your answer.**

8 **What are the advantages and disadvantages of using computer models to work out population sizes?**

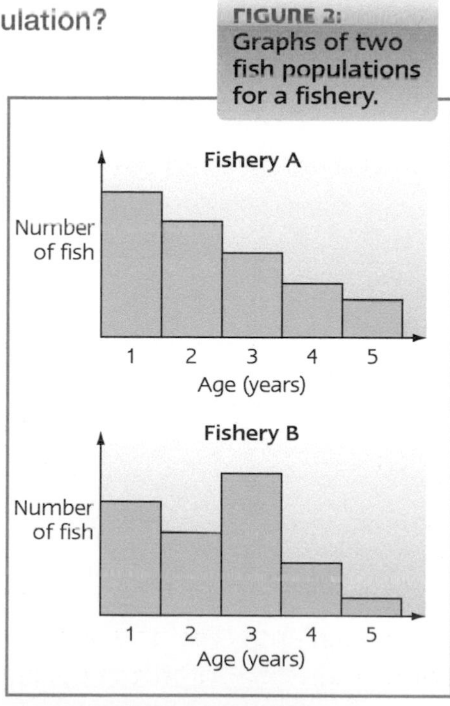

FIGURE 2: Graphs of two fish populations for a fishery.

Managing a Game Reserve

The rhinoceros and elephant are part of the Big Five animals in Africa. Their numbers have been threatened by poachers who are after the ivory from elephants' tusks and the rhinoceros' horns. Game reserves were set up to protect them. The Kruger game reserve in South Africa had to act to control the elephant numbers as they were threatening to destroy the habitat for other animals. They decided to cull (kill) selected elephants and over 14 000 elephants were culled between the years 1967 and 1994. In 1994 they stopped culling due to international pressure.

The number of elephants in the Kruger Park has now risen to over 14 000 and is growing at between 6 and 7% per year. By 2020 it is expected to be over 30 000 if nothing is done to manage the elephant population. The Kruger Park has said the park cannot support this number of elephants and the elephants left will become unhealthy through lack of food.

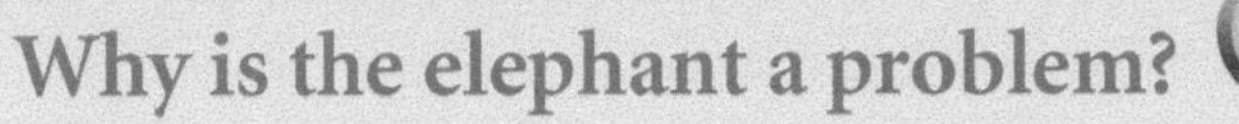

Why is the elephant a problem?

The elephant reaches sexual maturity between the ages of 9 and 12 years and it can continue to reproduce until 55–60 years of age and live even longer. The gestation period is 22 months and on average a female elephant gives birth every five years. The baby elephant weighs about 120 kg when it is born.

Elephants are herbivores and eat mainly grass but also twigs, roots, bark and fruit. They will eat for 16 hours a day consuming 140–270 kg of food per day. The elephant only digests 40% of its food, so dung beetles and termites thrive on the faeces (dung). Up to 16 000 dung beetles have been found in 1 kg of dung. The elephant uses its tusks to dig up roots, push bushes and trees out of its way, tear bark off trees and dig in dry river beds for water and salts. They also use their tusks for defending themselves and, in the case of the dominant male, to establish itself in the herd.

The elephant has no major predators. Lions will attack young elephants but few are killed. They move around the park in small groups of 10–20 elephants, trampling the bush as they go. Even though their skin is very tough, they need to take mud baths to cool down in the intense heat. Water is, therefore, essential for their survival. The ears also provide a large area to lose heat from.

What can be done?

The chief warden at the reserve must decide which strategies will reduce the numbers of elephants so they do not destroy the ecosystem.

The options are:

Culling

Elephants are killed. This must be done carefully as elephants grieve and show stress when members of the herd (especially young ones) are killed. In some cases a whole herd is destroyed. Culling does reduce the numbers to the required level.

Relocation

The elephant is captured by drugging it to put it to sleep. The elephant is then transported to another reserve where numbers are low. Only a few elephants can be moved and large bull elephants are never taken as they will upset the social balance in the new reserve. Fewer places are now available for relocation.

Contraception

Trials have successfully taken place where the female is injected with a contraceptive. The method has been shown to prevent the birth of new elephants but it does not reduce the numbers of elephants already in the reserve. It also requires tracking the elephants down to give boosters of the contraceptives – this is costly.

Extend the Game Reserve

The reserve may need to be increased by up to half its size. This will use the farming and housing land of the local population and will therefore increase poverty in the area.

Assess Yourself

1. What evidence is there that the elephant is a herbivore?
2. Produce a simple food web for the elephant and its dung.
3. How is the elephant adapted for:
   a obtaining food and water
   b protecting itself from predators
   c avoiding overheating.
4. Why have the numbers of elephants increased in the game reserve?
5. How are the elephants upsetting the balance of the ecosystem in the Kruger Park?
6. Explain why the Kruger game reserve resorted to culling elephants between 1976 and 1994.
7. Explain why there was an outcry against culling.
8. You are in charge of the game reserve and must work out short- and long-term plans to control the elephant numbers. Give a reasoned explanation of the strategies you will use.

ICT Activity

For one of the following animal species, prepare a scientific poster to indicate why they are in danger of extinction and what strategies are being used to protect them.

Gorilla   Bengal Tiger   Panda
Red Squirrel   Blue Whale

Maths Activity

Work out the amount of food an elephant consumes in a week, then in a year. Work out how much food would be consumed by a herd of 20 elephants in a year.

Work out how much food would be needed for the 30 000 elephants the Kruger Park predicts it will have by 2020.

Level Booster

8 Your answers show that you are able to present a reasoned explanation of an appropriate strategy to manage the elephant population.

7 Your answers show that you are able to explain the advantages and disadvantages of using contraceptives, culling and relocation.

6 Your answers show that you can explain why the elephant numbers have increased in the game reserve and why culling was used.

5 Your answers show that you are able to explain how the elephant is adapted for survival.

4 Your answers show that you are able to construct a simple food web for the elephant.

Recycling by rotters

BIG IDEAS

You are learning to:

- Describe the role of microbes in recycling
- Explain how the rate of microbe action can be increased
- Explain the composition of soil

Leaf litter

Dead leaves are found on the woodland floor. They form **leaf litter**. The leaf litter will also be the home of many animals as it gives them shelter and food.

Decay 

The amount of leaf litter does not keep increasing. The leaf litter is rotted down by microbes. Dead bodies of animals are also rotted down. This process is called **decay**.

1. Name **two** changes occurring to a leaf when it decays.
2. What might happen to the amount of leaf litter if no decay occurred?

Role of microbes in decay

The microbes that carry out most of the decay are fungi and bacteria – they are called **decomposers**. If you walk through woodlands in autumn, you will see a lot of fungi growing on dead wood. The conditions here are ideal for the microbes to grow. It is warm and moist with plenty of dead material available for the microbes.

3. Explain why little decay occurs in a desert.

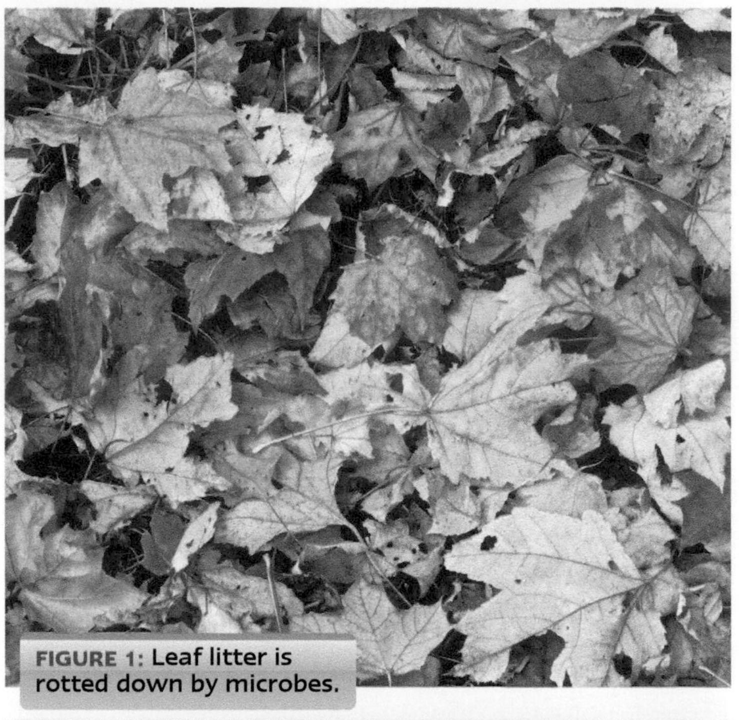

FIGURE 1: Leaf litter is rotted down by microbes.

FIGURE 2: Fungi are the most efficient organisms at breaking down wood.

Compost

Compost heaps are kept by gardeners to recycle garden waste. Grass, leaves, dead plants and vegetable remains are biodegradable and can be put on compost heaps. The heaps are regularly turned. The compost made can be added to the garden in the following year. This helps to put **nutrients** back in the soil.

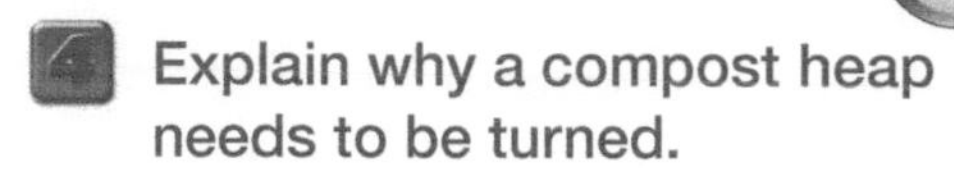

4 Explain why a compost heap needs to be turned.

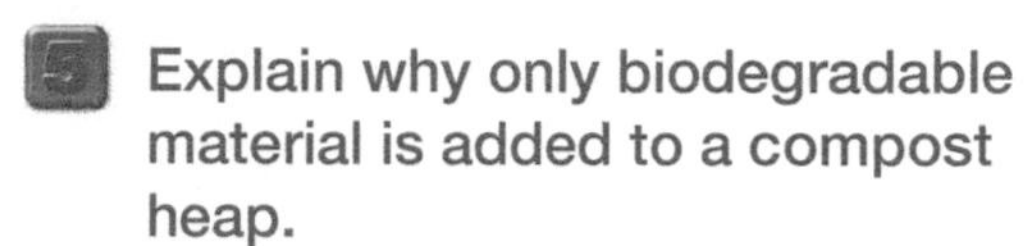

5 Explain why only biodegradable material is added to a compost heap.

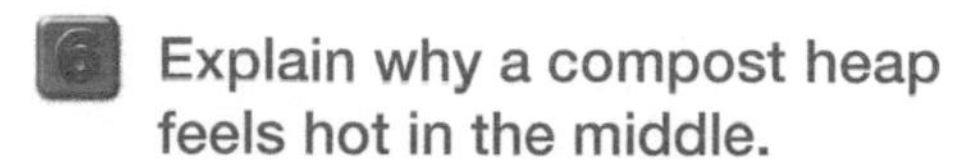

6 Explain why a compost heap feels hot in the middle.

FIGURE 3: Why are compost heaps so important to gardeners?

Did You Know...?

Kew Gardens has one of the biggest compost heaps in the world.

Forming soil

Decay is very important in the recycling of nutrients. Leaf litter is broken down to release minerals for plants to use. To do this the microbes carry out respiration and produce carbon dioxide and water and heat. The dead material forms a material called humus, which mixes with small particles of rock to form soil. This gives soil its dark colour and crumbly feel. Soils rich in humus hold more water and have more nutrients for plants.

How Science Works

It has been estimated that there is 105 kg of botanical waste per person per year. The council has decided to use two main strategies to recycle garden waste. The first is to provide compost bins to put the waste in. The contents of these are collected regularly and converted into mulches and compost for selling. The second approach is to sell tough plastic compost bins at reduced prices for the waste to be placed in. The compost formed can be removed and used to dig the garden. The containers have vents in and a flap at the bottom to remove the compost.

- Why are there vents in the compost bins?
- Why is the compost removed from the bottom not the top?
- What is the advantage to the council of its two strategies to recycle garden material?
- Why are the bins made from plastic, not wood or metal?
- It is recommended that the compost is dug into the garden in the winter. Explain why.

HSW

7 Explain why desert soils have little humus.

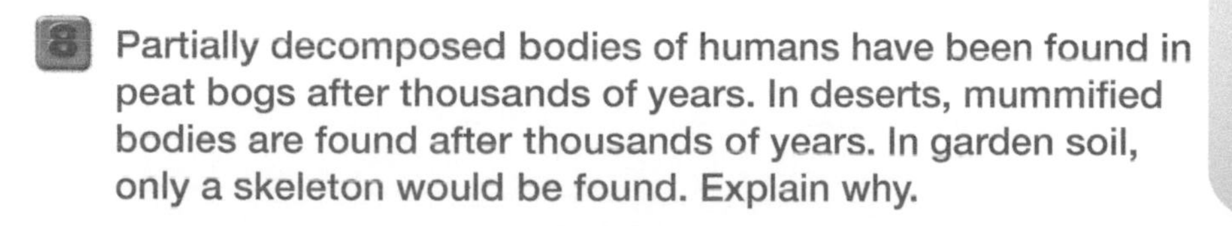

8 Partially decomposed bodies of humans have been found in peat bogs after thousands of years. In deserts, mummified bodies are found after thousands of years. In garden soil, only a skeleton would be found. Explain why.

How Science Works

Biodegradable plastics can be broken down by microbes. Biodegradable plastics must have at least 60–90% of their material decomposed in 60–180 days.

Populations

BIG IDEAS

You are learning to:

- Interpret predator–prey graphs (HSW)
- Explain how one organism can affect the population size of another organism
- Explain why organisms overproduce

5 Controlling the population

The number of organisms of one type (species) in an area is its **population**. Over many years the population size remains more or less constant.

A starling produces several chicks and each of them demands food. There are at least three chicks in the nest but not all of them will reach adulthood. The animals that survive are the strongest and healthiest. It is important to the **habitat** that there is no drastic increase or decrease in starling numbers.

FIGURE 1: Which of these chicks is most likely to survive to adulthood?

1. What could happen if too many starlings survived?
2. Starlings can regulate their population size. Why do they produce fewer eggs when there is little food?

6 Balancing numbers in a food web

It is important that the number of animals of a species does not increase rapidly and affect the other animals in a food web. An increase in numbers of one animal is usually balanced by an increase or reduction in numbers of another animal. Study the food web.

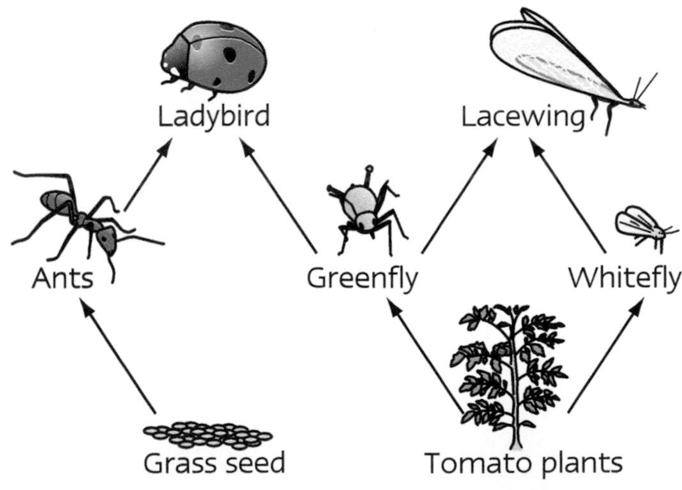

FIGURE 2: This food web consists of several linked food chains. Can you list the food chains?

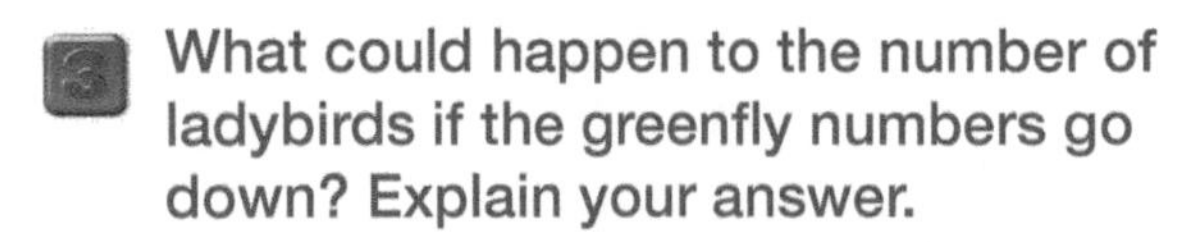

3. What could happen to the number of ladybirds if the greenfly numbers go down? Explain your answer.
4. What could happen to the number of whitefly if the greenfly numbers go down? Explain your answer.
5. Suggest what happens to the number of lacewings if the ladybird numbers increase. Explain your answer.

... habitat ... population

6 Yellowstone Park decided to open the park to the public. They wanted to make it safe and removed the bears, mountain lions and wolves from many areas.

a Explain why there was a rapid increase in the deer population.

b Explain why, after many years, few young deer were born and the deer became unhealthy.

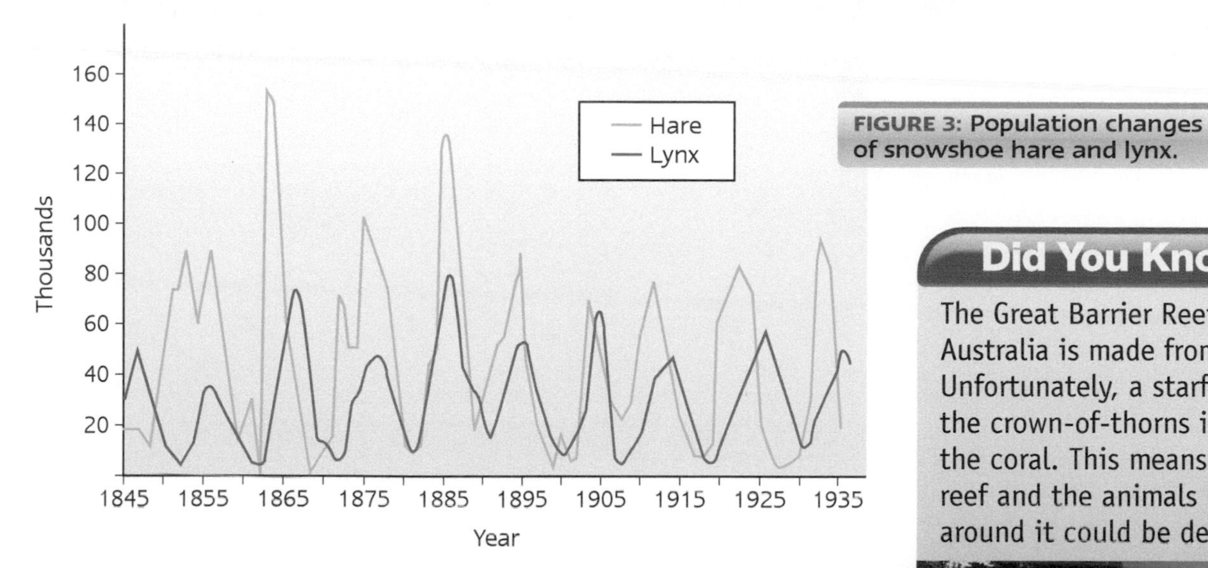

FIGURE 3: Population changes of snowshoe hare and lynx.

Did You Know...?

The Great Barrier Reef in Australia is made from coral. Unfortunately, a starfish called the crown-of-thorns is eating up the coral. This means the whole reef and the animals living around it could be destroyed.

Predator – prey

The graph above shows the changes in the numbers of lynx and snowshoe hare in an area in Canada over 90 years. The numbers of snowshoe hare increase, as there is more food available. This means the numbers of lynx increase as they have more food. The number of **prey** then falls followed by a fall in **predators**.

7 Describe how the patterns for the populations of snowshoe hare and lynx compare.

8 Give evidence to support the claims that:

a The lynx control the hare population.

b The hare population size controls the number of lynx.

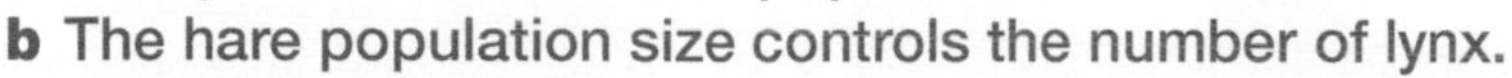

FIGURE 4: The snowshoe hare in spring.

FIGURE 5: The lynx.

Biological control

BIG IDEAS

You are learning to:
- Explain the benefits and problems caused by biological control and pesticides
- Interpret graphs on biological control HSW
- Explain what is meant by a pesticide

Pests all around us

Pests destroy garden plants, trees and feed off animals.

Greenfly are pests on roses, whitefly on tomato plants, slugs on lettuces and ticks and lice on sheep. Humans have tried to get rid of pests by using chemicals sprays.

1. What is a pest?
2. Why do you get lots of slugs when you grow lettuces?
3. Why are chemicals used to control pests?

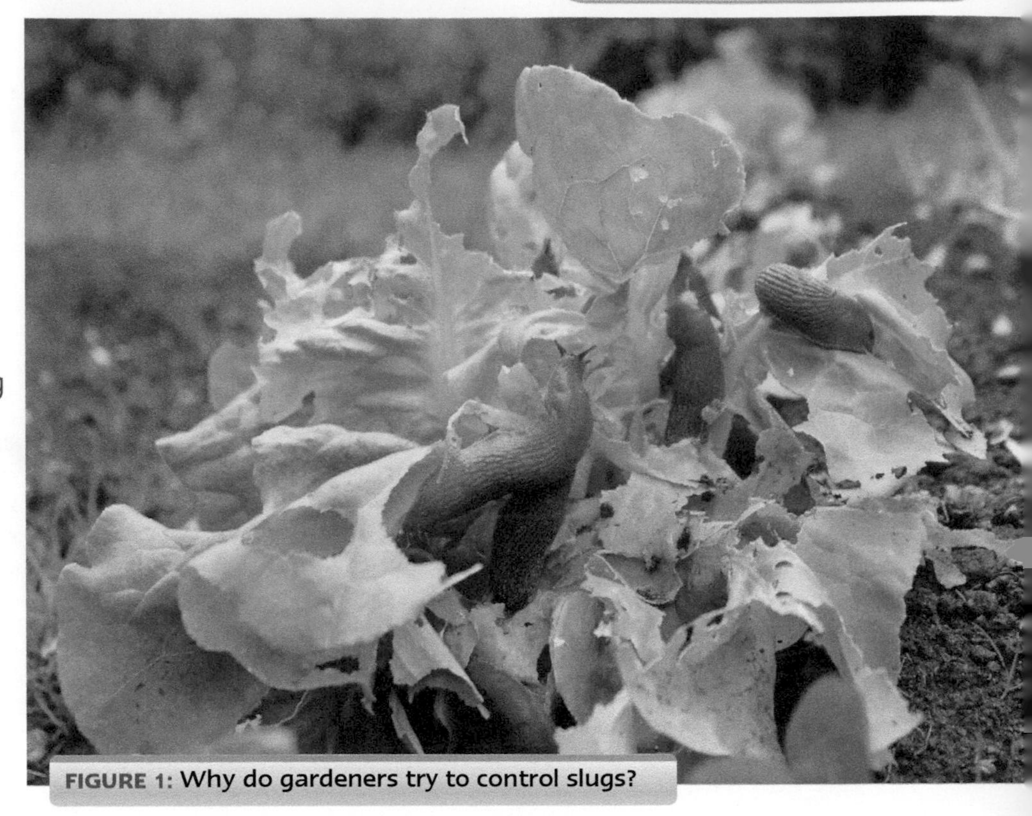

FIGURE 1: Why do gardeners try to control slugs?

Pesticides

The chemicals are called **pesticides**. They kill the pests quickly once they are used. There are several types of pesticide: herbicides kill weeds; insecticides kill insects, e.g. greenfly, and fungicides kill fungi.

4. Why is it important that fungicides do not kill insects?
5. The greens for a golf course must only have grass on them. Name a weed that might be found on them.

Pesticides or predators

Many problems have come from the use of pesticides.

- Early pesticides were poisonous to other animals so useful insects, like bees, were also killed by the poisons.
- The pesticides built up in the environment because they did not break down. This led to the death of many carnivores at the top of the food chain.
- Some pests became resistant to pesticides, so higher concentrations had to be used. It is now difficult to kill off mosquitoes with a pesticide called DDT.

Many of the pesticides being produced nowadays have less effect on other animals and do break down. **Biological control** uses natural means to control pests, for example using predators like ladybirds to eat greenfly and whitefly.

6 Name **one** advantage of using pesticides.

7 Name **one** advantage of using predators.

FIGURE 2: Ladybirds are loved by gardeners. Do you know why?

How Science Works

Biological control doesn't always work. The cane toad was introduced into Australia to control the cane beetle; it is now a major pest. It eats many of the native animals and has few predators.

Did You Know...?

Cats were introduced into Baltimore to control the rats. Unfortunately, instead of eating rats they raided the dustbins.

Biological control

Whitefly have started to attack the tomato plants in a greenhouse. Ladybirds are added to the greenhouse to control the number of whitefly. The graph shows the changes in population in whitefly and ladybirds over a few months.

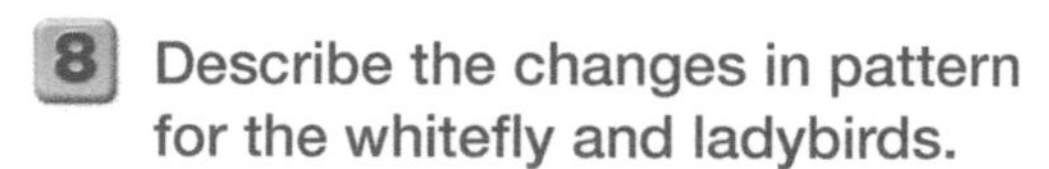

8 Describe the changes in pattern for the whitefly and ladybirds.

9 Explain why these changes occur.

10 If the ladybirds died out after five months because there are few whitefly left, predict what would happen to the number of whitefly and explain why.

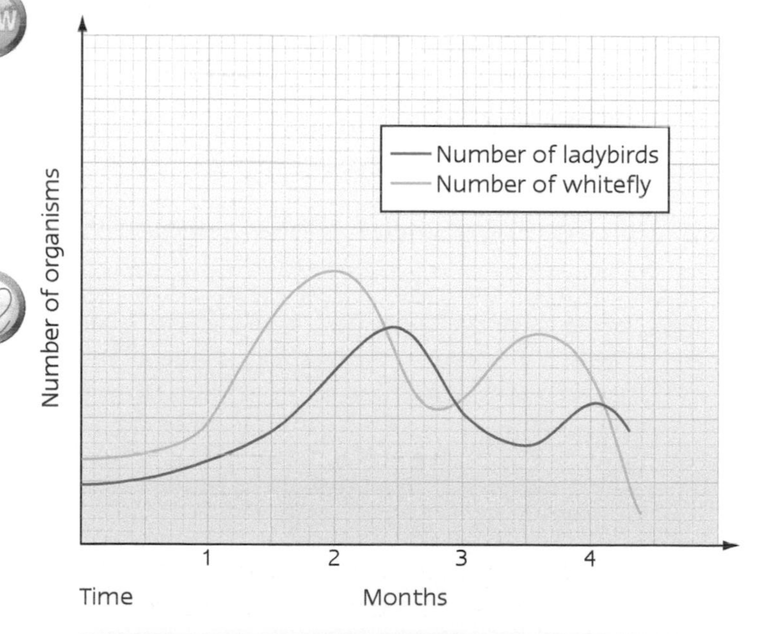

FIGURE 3: Graph to show the population change in whitefly when ladybirds are introduced.

Practice questions

1 Copy the following passage into your exercise book and complete it using the words below:

predators **population** **competition**

The number of rabbits in an area is the ______ of rabbits. The numbers will reduce if ______ like the fox eat the rabbit or there is ______ for food between the rabbits.

2 Owls eat mice.

a Give **two** ways the owl is adapted to catch mice.

b Why does the owl catch several mice?

c What could happen to the owls if pesticides killed the mice?

3 Grass cuttings are put on a compost heap. After a while they start to rot.

a Which microbes cause the rotting?

b The compost heap is turned to let air in. Which gas do the microbes need?

c Explain why less decay occurs in frosty conditions.

d Why are plastic bags not put in compost heaps?

4 Herds of zebra are found on the grassland areas of Africa. They are hunted by lions who work in small groups to creep up on the herds. The lions kill very few zebra.

a Which zebra are the most likely ones to be killed?

b How is the zebra adapted to avoid being killed by the lions?

c The lion eats few zebra. What other factor controls the number of zebra in the area?

d The zebra leaves a lot of dung on the grassland area. What happens to it?

5 Scientists were studying the growth rate of yeast when wine was being made. They set up a flask containing a glucose solution and added yeast.

Over the next few days they took samples to count the number of cells.

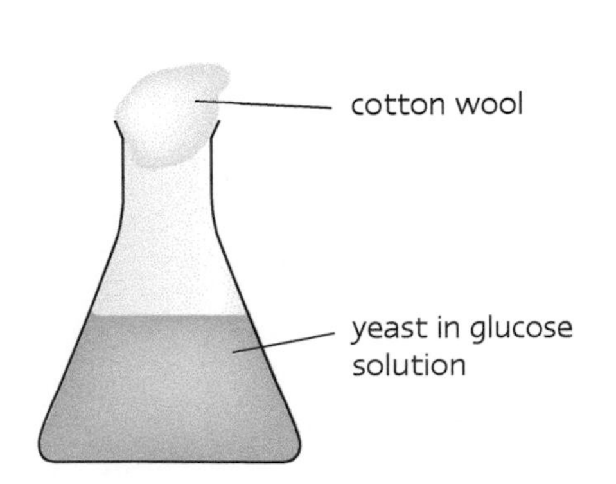

a Plot a graph of the results.

b Which is the independent variable?

c Which is the dependent variable?

d During which period of time did the yeast grow quickest?

e Why has the number of yeast cells levelled off after seven days?

Time (days)	Number of yeast cells/mm^3
1	25
2	80
3	200
4	400
5	550
6	600
7	650
8	650

6 Scientists were studying the growth of salmon in a fishery. They found that not all the food that was given to the salmon was used for new growth.

a Explain what happens to the chemical energy in food not used for new growth.

b Explain why in nature food chains have a maximum of five links.

The scientists found a parasite fungi was feeding off the salmon. They decided to use a pesticide.

c What is the advantage of using a pesticide?

d What problems might it cause?

7 The lynx eats snowshoe hare. Over many years, the population sizes of lynx and snowshoe hare were estimated in Canada by counting the skins brought in by hunters.

a Plot a graph of the results shown in the table.

b In which period are the snowshoe hare numbers greatest?

c Explain why the numbers of snowshoe hare go up and down.

d Suggest how other species may be affected by the changing population of snowshoe hare.

e How effective a way is this of monitoring populations?

Period	Snowshoe hare skins brought in (thousands)	Lynx skins brought in (thousands)
I	20	10
2	55	15
3	65	55
4	95	60
5	55	20
6	5	15
7	15	10
8	50	60
9	75	60
10	20	10
11	25	5
12	50	25
13	70	40
14	30	25
15	15	5

Learning Checklist

4

- ☆ I know the conditions that affect a population size. page 170
- ☆ I know the differences between a predator and prey. page 177
- ☆ I can explain what a pest is. page 178
- ☆ I know what pesticides are. page 178

5

- ☆ I can identify the main adaptations a predator has to survive. page 166
- ☆ I know that decomposers break down dead material. page 174
- ☆ I know bacteria and fungi are decomposers. page 174
- ☆ I know the conditions needed for dead material to rot. page 174
- ☆ I can explain what happens to the organisms in a food web if one organism disappears. page 176

6

- ☆ I can explain why not all the energy is passed on from a predator to a prey. page 168
- ☆ I can explain the advantages and disadvantages of using pesticides. page 178

7

- ☆ I can construct pyramids of numbers for parasitic food chains. page 169
- ☆ I can interpret complex predator–prey graphs. page 177
- ☆ I can explain why food chains usually have a maximum of five links. page 181

8

- ☆ I can construct pyramids of energy. page 169
- ☆ I can explain the effect of the accumulation of pesticides on a food chain. page 178

Topic Quiz

1 Name **two** types of microbe involved in the decay of leaves.

2 What is a decomposer?

3 Why are toadstools most commonly found in woodlands in the autumn?

4 Rabbits reproduce rapidly. Give **two** reasons why their numbers stay fairly constant.

5 Not all the food energy eaten by a mouse is used for new growth. Name **two** reasons for this.

6 Why are many woodlice found in compost heaps?

7 What shape is the pyramid of numbers for the food chain:

Grass → **Rabbit** → **Flea**

8 Why are there very few foxes in an area of woodland?

9 What is the benefit of biodegradable pesticides?

True or False?

If a statement is false then rewrite it so it is correct.

1 All bacteria cause disease.

2 A population is the number of all the animals in an area.

3 All predators are carnivores and all prey are herbivores.

4 Fungicides are fungi used to kill pests.

5 Predation is important to a prey as it allows only the strongest ones to survive.

6 Compost heaps break down to form the humus part of soil.

7 One of the reasons many worms are found in compost heaps is that they are very moist.

8 Energy is lost from a food chain as heat.

Literacy Activity

Giant tortoises are one of the native species found on the Galapagos Islands. They can live for 150 years. They feed on grass and small shrubs. They are in danger of extinction. The Judas goat was introduced to the Islands by whalers and allowed to roam wild. They are eating up the vegetation and causing soil erosion on the mountain side. Environmentalists have decided they need to be removed but the goats can reach inaccessible parts of the island. Helicopters have been used to find the goats so they can be shot. Once the goat numbers are reduced they will be rounded up and removed from the Islands.

1 What is meant by the native species?

2 Why is the Judas goat not a native species?

3 Why are they hunting the goats with helicopters?

4 Explain whether or not you think they are right to cull the goats.

ICT activity

Present a PowerPoint display of a biological control used by man which has not worked. Describe the problem that it was used for. Explain what went wrong and why it went wrong. Indicate if other solutions were being used.

Visitations?

By the end of this unit you will be able to describe how the planets are arranged around the Sun and move in orbits. You will be able to explain how the force of gravity keeps the planets in orbit and causes the effect of weight. You will be able to describe how scientific ideas have affected society.

For thousands of years people have claimed to see lights and strange unexplained objects in the sky. Many people believed, and continue to believe, that we have been visited by living creatures from outside our own Solar System. There have been countless photographs taken of objects that are supposed to be UFOs or Unidentified Flying Objects. Any internet search for 'UFOs' will turn up many of these images. Most are relatively easy to explain. They may be as a result of aircraft or unusual cloud formations. The vast majority are obviously fakes but some seem to defy detailed investigation.

FIGURE 1: Typical picture of a flying saucer.

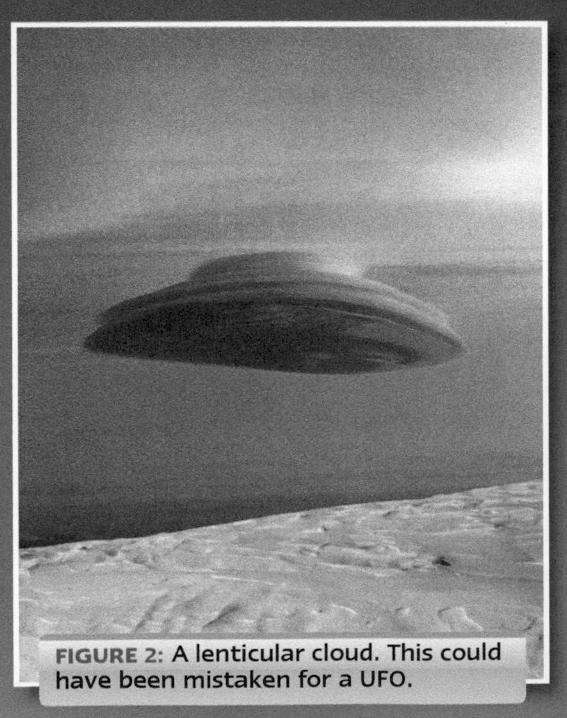

FIGURE 2: A lenticular cloud. This could have been mistaken for a UFO.

Of course, film makers have for many years made films involving visits from alien life forms and created many impressive books and films as a result. Films such as *Star Wars*, *Star Trek* and *Alien* are about how people respond to these challenges.

Indeed, as the Universe is so vast the chances are that somewhere 'out there' there are other life forms and that we are not alone in the Universe. But, as yet, there are no confirmed sightings of, or meetings with, creatures from another world. Any alien life that did visit the Earth would almost certainly be more advanced than our own. They would have to possess the technology necessary to travel vast distances to get here.

One of the most famous incidents took place at Roswell in New Mexico in 1947. On the evening of 2 July 1947, many UFO sightings were reported during a thunderstorm near Roswell. Some reports indicated that an object had been struck by lightning. The next day, strange wreckage was found in a field and, when the impact site was located, a UFO craft and alien bodies were allegedly found and an autopsy conducted. On 8 July 1947, the *Roswell Daily Record* announced the capture of a flying saucer. But if it existed no one knows what happened next, or if they do then they are not saying!

The official explanation was that the 'UFO' was a weather balloon. Many people, including the residents of the town, thought that there was more to it than that.

In the following days virtually every witness to the whole event was either quickly transferred or seemed to disappear from the face of the Earth. This led to suspicions that an extraordinary event was the subject of a deliberate government cover up. Over the years, books, interviews and articles from a number of military personnel who had been involved with the incident, have added to the suspicion of a deliberate cover up.

In Roswell itself there is a museum which claims to show a model of the alien that landed at Roswell.

FIGURE 3: The 'Roswell Incident'.

FIGURE 4: A model of the alien from Roswell.

What do you know?

1 What is a UFO?
2 What is the furthest that man has travelled into space?
3 Which of the planets in our Solar System can you name?
4 What is the difference between a star and a planet?
5 Why do we have day and night?
6 Why do astronauts need to wear space suits?
7 Do you think that any human structure or building can be seen from space? If so, which ones?
8 Can you think of any uses for satellites?
9 What is meant by a 'cover up'?
10 Why do you think that the incident at Roswell, if it ever occurred, was covered up?
11 Why is there a chance that there is alien life somewhere in the Universe?
12 Suggest why scientists have sent radio messages into space.

BIG IDEAS

You are learning to:

- Discuss why we have day and night
- Explain what determines a year
- Understand how scientific ideas about the Solar System have developed 

The spinning Earth

We all live on the **Earth**. It is a sphere so it is like a big, round ball. The Earth gets light and heat from the Sun and this enables living things to survive on it.

The Earth is always spinning round, or **rotating** on its axis. This can be hard to believe because it does not feel as if we are moving. This is because we are also moving at the same speed and so we do not feel any motion. The Earth always spins at the same constant speed (about 1000 miles per hour at the equator!) and it takes about 24 hours to make one complete spin or rotation.

FIGURE 1: Earthrise taken from the Apollo 8 spacecraft in 1968. On the Earth, the sunset terminator crosses Africa.

Nothing flies off the planet even though it is travelling so quickly because the pull of **gravity** holds everything in place on the surface.

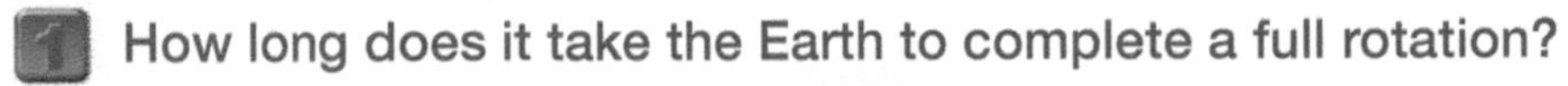

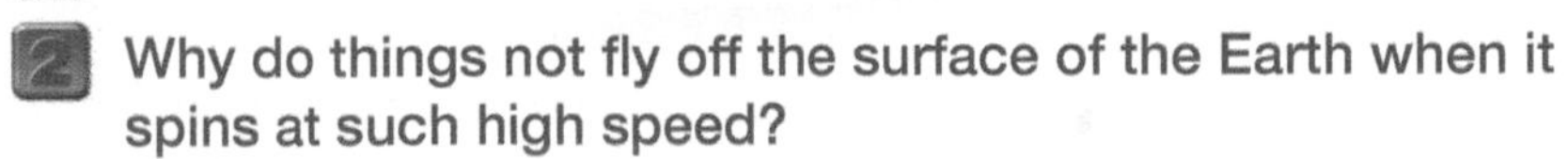

How Science Works

It was only once satellites were able to take photographs of the Earth from space that it was possible to finally prove that the Earth is spherical. Until then some people still thought that the Earth was a flat disc.

Day and night  

The Sun seems to move across the sky every day because it rises in the East and sets in the West. For thousands of years people thought that this was what happened.

Eventually, in 1543, a Polish priest called Nicolaus Copernicus realised that it was the Earth that moved around the Sun. We now know that it takes about a year for the Earth to complete one full orbit. It actually takes $365\frac{1}{4}$ days to complete one orbit. So every four years there is a leap year where there is an extra day in the calendar.

Light from the Sun falls on one half of the spinning Earth. This half of the Earth is in **daylight**. The other half of the Earth does not receive any sunlight so it is **night** here. As the Earth spins, we move through the light, into the darkness, and back again. Because the Earth is always spinning at the same rate we see a pattern of day and night. Half of the Earth is always in the light and half of the Earth is always in the dark.

... daylight ... Earth ... gravity

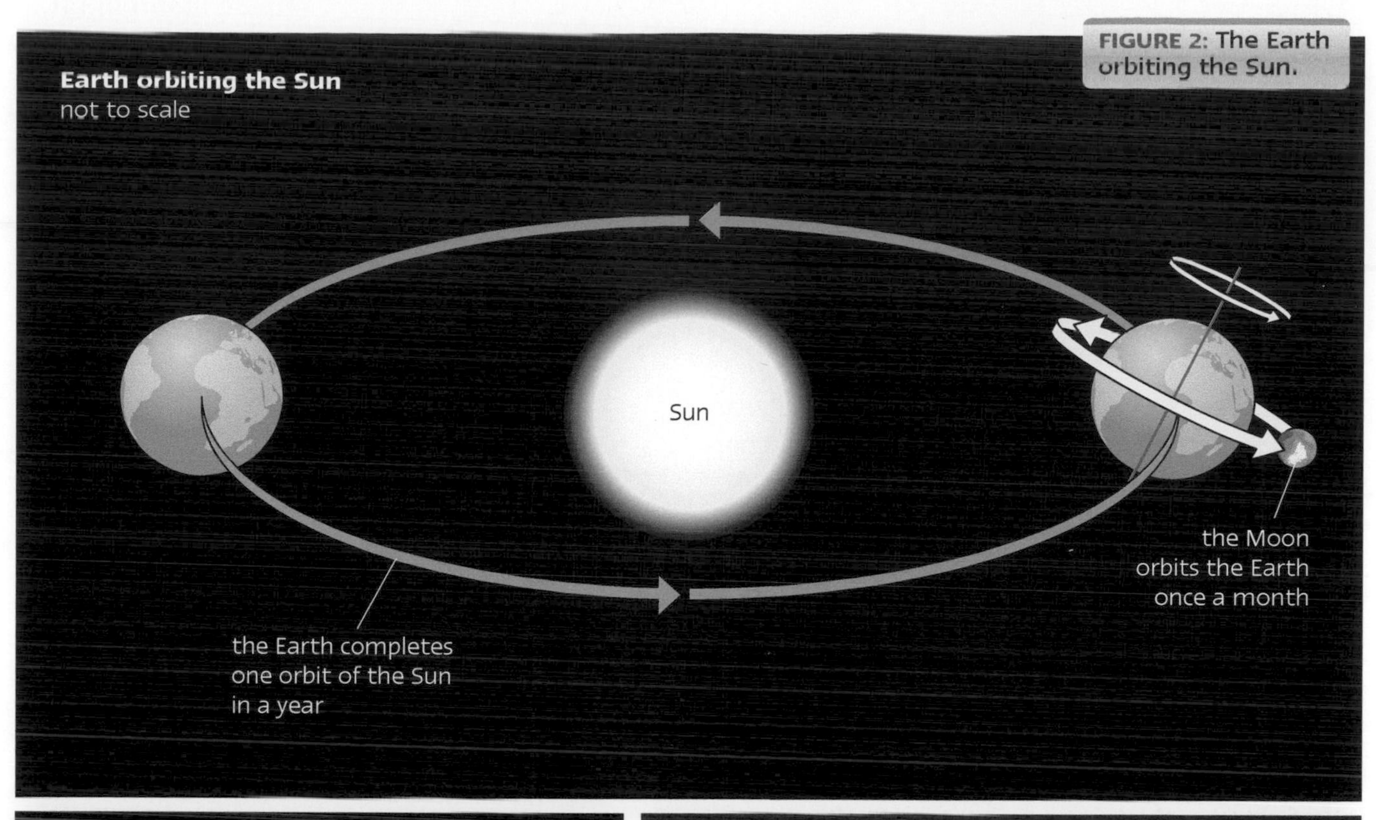

FIGURE 2: The Earth orbiting the Sun.

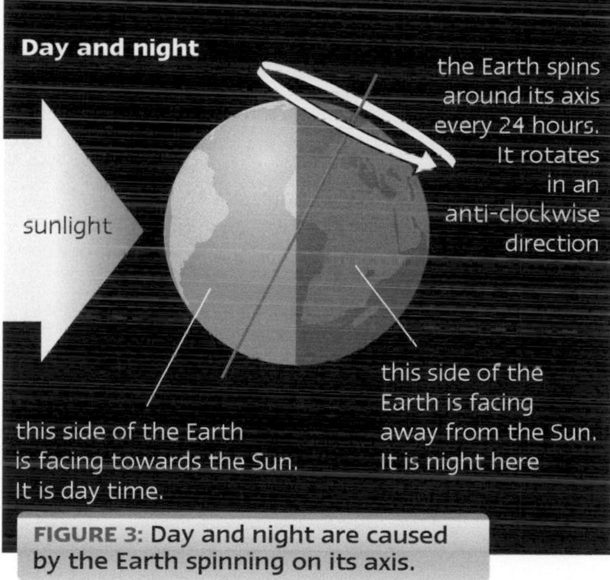

FIGURE 3: Day and night are caused by the Earth spinning on its axis.

FIGURE 4: Satellite image of Europe at night.

3. Why do we get day and night?
4. Why do we have 'leap years'?
5. Explain why people used to think that the Sun moved around the Earth.
6. What evidence would be needed to prove that the Earth does not orbit the Sun? Explain your answer.

... *night* ... *rotating*

The seasons

BIG IDEAS

You are learning to:
- Explain the causes of seasonal variation during the year
- Explain the causes of seasonal variation between the Northern and Southern hemispheres

The seasons

The weather in the UK varies a lot during the year and so does the amount of daylight. The weather varies according to the **season** of the year. The weather is linked to the seasons. The seasons happen because the Earth's axis is **tilted**.

The Earth has two **hemispheres** or halves. The equator is an imaginary line around the centre of the Earth that separates the two hemispheres. When the Northern hemisphere is tilted towards the Sun then it is **summer** in the Northern hemisphere and **winter** in the Southern hemisphere. When the Northern hemisphere is tilted away from the Sun then it is winter in the Northern hemisphere and summer in the Southern hemisphere.

In summer the Sun is higher in the sky and the days are longer than the nights. In summer it is warmer because the Sun is shining for longer and it warms up the Earth more. In winter the Sun is lower in the sky and the nights are longer than the days. It is colder because the Sun does not heat up the Earth as much.

At the spring and autumn equinoxes, day and night are of equal length. The longest day is called the summer solstice. It occurs on 21 June in the Northern hemisphere. On this day the Sun is at its highest position in the sky. The winter solstice occurs on 21 December. The sun is at its lowest position in the sky and the day is the shortest one in the year.

FIGURE 1: Same place in summer and winter showing the difference in the weather.

1. Why is it warmer in the summer than it is in the winter?
2. On which date would the summer solstice occur in the Southern hemisphere?
3. Explain why the seasons are less apparent to people who live on or near to the equator.
4. Why do you think plants do not grow as much in the winter as they do in the summer? Explain your answer.

... hemisphere ... season ... summer

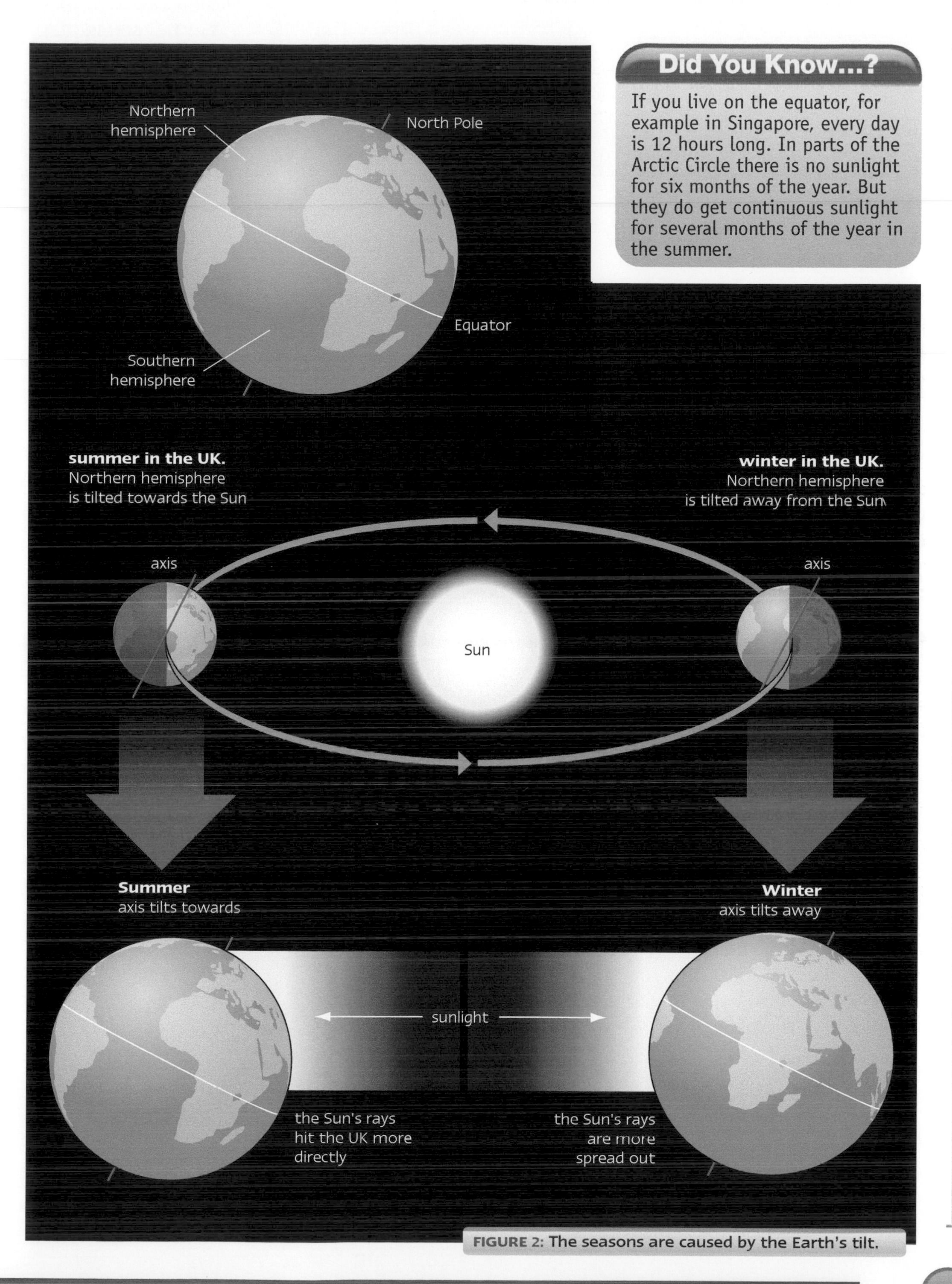

FIGURE 2: The seasons are caused by the Earth's tilt.

Did You Know...?

If you live on the equator, for example in Singapore, every day is 12 hours long. In parts of the Arctic Circle there is no sunlight for six months of the year. But they do get continuous sunlight for several months of the year in the summer.

The Moon

BIG IDEAS

You are learning to:

- Recognise that the Moon is a natural satellite of the Earth
- Understand why the shape of the Moon appears to change during its orbit of the Earth
- Understand what causes solar and lunar eclipses

The Moon

The Moon is the only natural object that **orbits** the Earth. Our Moon takes about 28 days to orbit the Earth. This period of time is known as a lunar month.

The Moon is the fifth largest moon in the Solar System. Most moons are much smaller than their planets. The Moon, however, is relatively large, about a quarter of the diameter of the Earth. It is about 384 000 km away from the Earth.

It does not have an atmosphere so there is no air, no water and no wind. As far as we know there is no life on our Moon.

The Moon is lit up because light from the Sun hits it. We see our Moon because it **reflects** this light into our eyes.

1. What is a lunar month?
2. Why do we see the Moon?

Did You Know...?

The first man to walk on the Moon was the American astronaut Neil Armstrong on 20 July 1969. Upon taking his first step onto the Moon, he said: 'That's one small step for a man, one giant leap for mankind.' But the word 'a' was lost in radio transmission. What do you think he meant by this?

FIGURE 1: The twelfth and last man to walk on the Moon was the American astronaut Harrison Schmidt in 1972. No one has returned to the Moon since.

Phases of the Moon

The Moon is an example of a natural **satellite**. A satellite is the name given to any object that orbits a larger object. The Moon is held in place around the Earth as a result of the Earth's **gravity**.

Because the Moon orbits the Earth, its appearance changes over the course of a month. One half of the Moon is always lit by the Sun but we cannot always see all of this lit part.

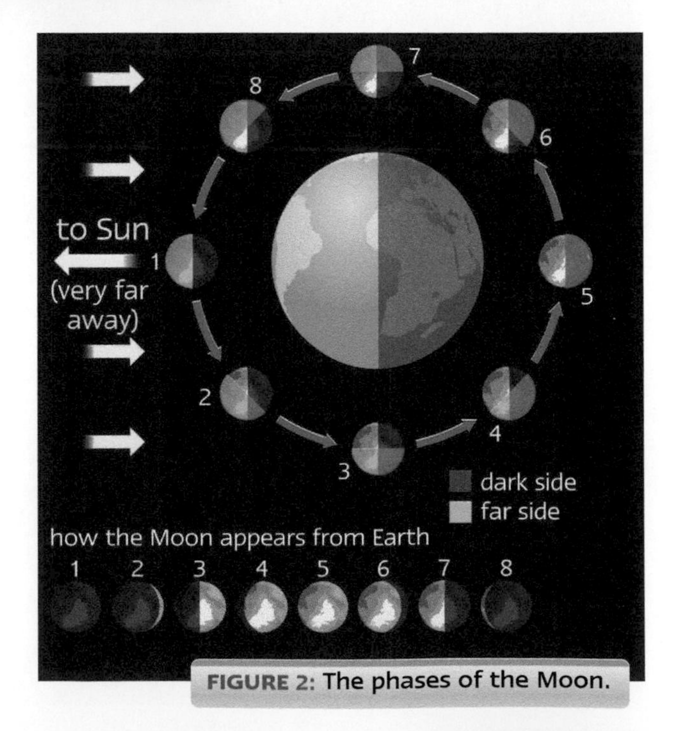

FIGURE 2: The phases of the Moon.

... *eclipse* ... *gravity* ... *lunar eclipse* ... *orbit*

In Figure 2, you can see the Moon always has a lit side (facing the Sun) and a dark side (facing away from the Sun). The lit side does not always face the Earth. From the Earth, we can only see the part of the Moon that is facing towards us at any time. This means that we only see part of the Moon at any one time.

The appearance of the Moon changes as it orbits the Earth. This change of appearance is called the **phases** of the Moon. The phases we see result from the angle the Moon makes with the Sun as viewed from the Earth.

When the Moon is between the Sun and the Earth we cannot see any of the lit part. This is called a new moon. When we see all of the lit side it is a full moon. During the lunar month, the Moon goes through all its phases. You can see the phases drawn in Figure 2.

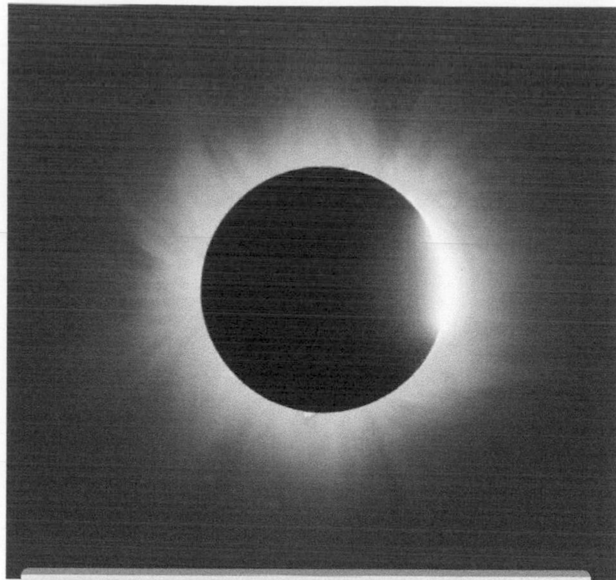

FIGURE 3: A solar eclipse showing the corona.

3 What is meant by the word 'satellite'?

4 Why does the appearance of the Moon seem to change during a month?

Eclipses

Sometimes when the Moon is between the Sun and the Earth some parts of the Earth are in the shadow of the Moon. People standing in the shadow see a **solar eclipse** where the sun is totally blocked out.

Sometimes the Moon moves into the shadow of the Earth. The Moon appears very dark. This is called a **lunar eclipse**.

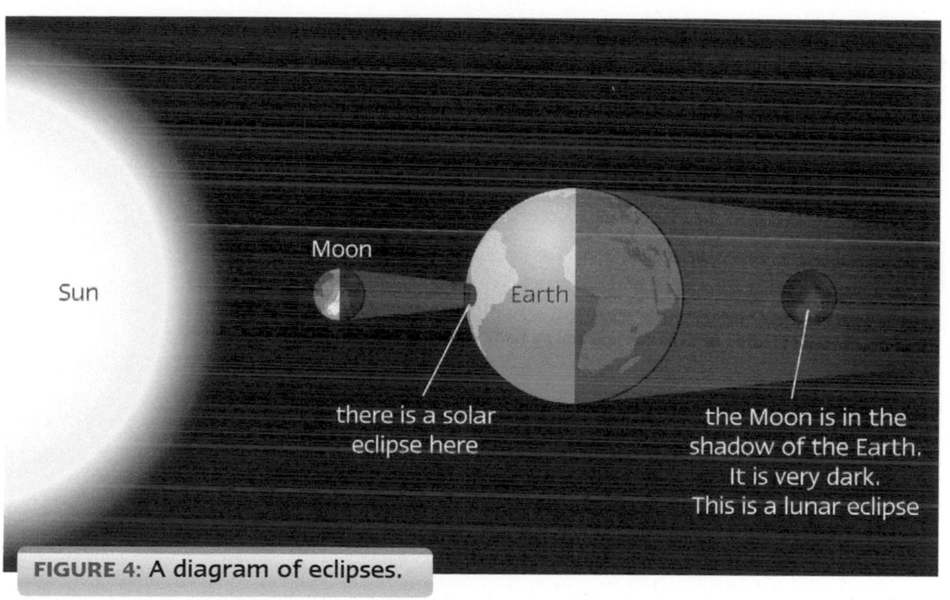

FIGURE 4: A diagram of eclipses.

5 What is the difference between a solar eclipse and a lunar eclipse?

6 Would a solar eclipse happen when there was a full moon or when there was a new moon? Explain your answer.

7 Explain as fully as you can why no one has returned to the Moon since 1972.

The Solar System

BIG IDEAS

You are learning to:
- Understand the range of objects in the Solar System
- Recognise the differences between planets and stars
- Identify patterns in the data about the planets 

Our Solar System

When we look up at the night sky we see countless billions of **stars**.

We cannot see them all with the naked eye. As **telescopes** have improved so we have been able to see more and more stars because we can look further and further into space.

Our **planet** – the Earth – is just one of the planets that travel around, or **orbit**, the Sun. The Solar System is the name given to our Sun and all the planets that are orbiting it.

 1 What is the Solar System?

 2 How has our knowledge of space improved over time?

Watch Out!

There were always thought to be nine planets in our Solar System but in 2006 Pluto was reclassified as a dwarf planet due to its small size and because it was very similar to the largest of the objects found in the asteroid belt.

The planets

Planets do not give out light. We see them because they reflect light and because they are much closer to us than the other stars in the sky. All the planets are kept in orbit by the force of **gravity**.

Travelling outwards from the Sun, the order of the planets is Mercury, Venus, Earth, Mars, Jupiter, Saturn, Uranus and Neptune. You can use a sentence to help you to remember the order of the planets, like this one:

'My Very Easy Method Just Speeds Up Naming (planets)'

The closer the planet is to the Sun, the hotter it is. Apart from the Earth the other planets are either too hot or too cold for anything to survive on them.

The planets in the Solar System are split into two groups. The four planets closest to the Sun are rocky planets made from solid material. They are called the inner planets. The remaining planets are called the outer planets. They are mostly made from gas and are sometimes called gas giants. 

Most of the planets also have **moons** orbiting them. There are also thousands of **asteroids** which are small lumps of rock. Most of the asteroids are found in the asteroid belt which lies between Mars and Jupiter.

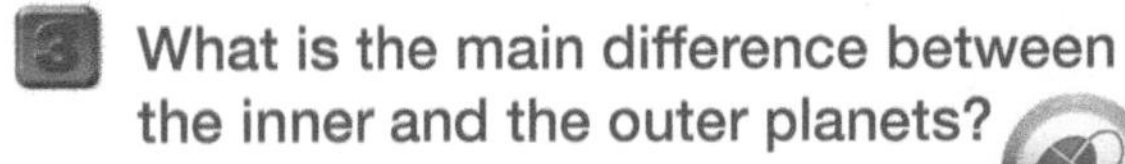
3 What is the main difference between the inner and the outer planets?

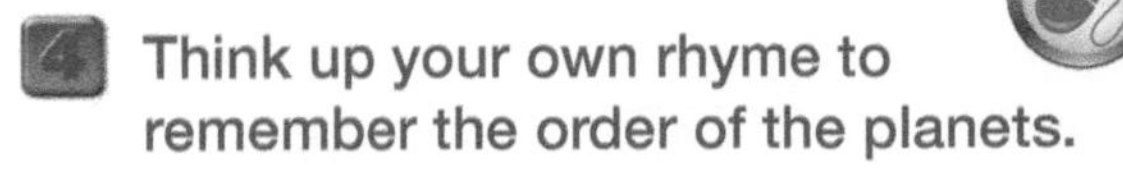
4 Think up your own rhyme to remember the order of the planets.

5 How do we see the planets in our Solar System?

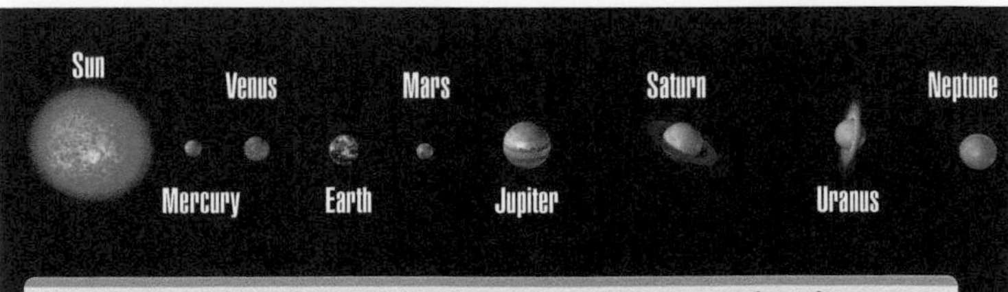

FIGURE 1: The planets in the Solar System. The diagram is not drawn to scale. The relative size of the planets and their relative distances apart are not correct.

The scale of things

It is very difficult to get an idea of the size of the Solar System because of the huge distances involved. For example, if the Sun was 15 metres in diameter then, to the same scale, the Earth would have a diameter of 1.5 centimetres and would be 100 metres away from the Sun.

The further the planet is from the Sun the longer it takes it to complete an orbit of the Sun. So Mercury orbits the Sun in 88 days while Neptune takes 160 years.

Here are some more facts about the Solar System.

	Mercury	Venus	Earth	Mars	Asteroids	Jupiter	Saturn	Uranus	Neptune
Diameter (km)	5000	12 000	12 800	7000	–	140 000	120 000	52 000	50 000
Distance from Sun (million km)	60	110	150	230	–	780	1400	2900	4500
Time to orbit the Sun (years)	0.2	0.6	1	2	–	12	30	84	160

6 Put the planets in order of size with the largest first.

7 Why does it take longer for Saturn to complete an orbit of the Sun than it does for Jupiter?

Did You Know...?

Olympus Mons (on Mars) is the largest volcanic mountain in the Solar System. It is 600 km across and 27 km high! And you thought Mount Everest (about 8.5 km high) was tall!

FIGURE 2: Olympus Mons.

The Sun

The Sun is at the centre of our Solar System. The Sun is a star which means that it produces its own heat and light as a result of a series of **nuclear** reactions in its core.

The Sun makes up 99.86 per cent of the Solar System's mass! That means that all the planets put together (including Jupiter) as well as all the asteroids only make up about 0.14 per cent of the mass of the Solar System.

Millions of stars form a **galaxy**. The stars in a galaxy are held together by the force of gravity between them. There are countless billions of galaxies in the **universe**.

8 How does the Sun produce heat and light?

9 Put the following in order of size, smallest first:
planet universe star asteroid Solar System moon galaxy

10 What sort of evidence have scientists used to develop their understanding of the Solar System?

11 Why has the amount and reliability of the evidence available to astronomers increased significantly over the last 50 years?

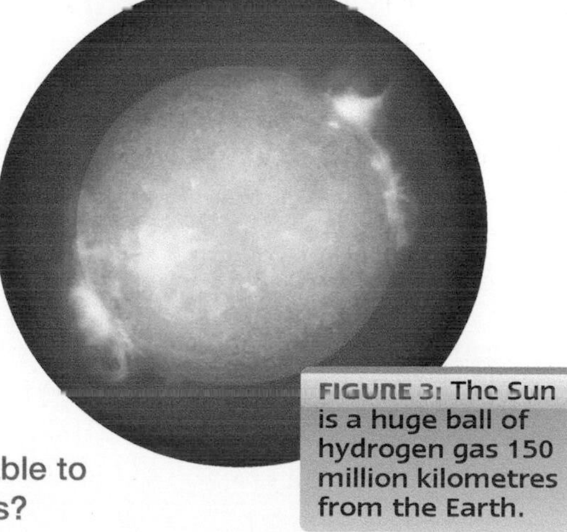

FIGURE 3: The Sun is a huge ball of hydrogen gas 150 million kilometres from the Earth.

Gravity in space

BIG IDEAS

You are learning to:
- Describe what gravity is and where it acts
- Explain the factors affecting the strength of gravity
- Recognise that gravity keeps planets and satellites in orbit

Gravity 

When any two objects are close together, there is a force of **gravity** which tries to pull them together. The more mass that the objects have, the stronger the pull of gravity between them.

The Earth is a large, **massive** object so the gravity that it produces is strong enough to be noticed. The gravity from the Earth holds you in place on the surface of the Earth and pulls you back down when you try to jump upwards.

The Sun is more than 300 000 times more massive than the Earth. Because it is so massive it produces a huge force of gravity. This extends out into space and keeps the planets in **orbit** around it. If the Sun did not produce this amount of gravity then the number of planets held in orbit around it would be less.

Isaac Newton (1642–1727) was the first person to try to understand what gravity is and what effect it has on objects. He did not discover gravity. He was the first person to explain how it works. Newton's work on gravitation is one of the most important scientific explanations of all time.

1 Why does the Earth produce enough gravity for us to feel it's force?

2 What would happen to the planets if the Sun's gravity was suddenly switched off?

FIGURE 1: Comets are kept in orbit around the Sun because of the Sun's gravity. They have elliptical orbits, not circular ones.

FIGURE 2: All these things are held in place on the surface of the Earth because of gravity pulling them down.

Did You Know...?

Newton is supposed to have thought of the law of gravitation when an apple fell on his head.

Newton explains gravity

Newton realised that the force that held the Moon in orbit round the Earth was the same force that made an apple fall from the tree onto the ground. This force is called gravity.

... artificial satellite ... distance

Gravity exists throughout the Universe. Its strength is always dependent on the mass of the two objects and their **distance** apart. The further apart the objects are, the weaker the force of gravity between them is. The more massive the objects are, the greater the force of gravity between them.

The gravity from the Earth keeps the Moon in orbit around it. The moons of the other planets in the Solar System are also held in place in their orbits by the strength of the pull of gravity from their nearest planet.

They do not fall into the planet because of their speed which moves them forward parallel to the planet's surface and the pull of gravity which acts downwards towards the planet. All orbiting objects stay in their orbits because of their speed and the force of gravity.

How Science Works

Newton's Universal Law of Gravitation has been used to explain why it is that the planets do not follow precise orbits but wobble along their path. This is due to the attraction of other planets. It was the irregular motion of Uranus that led to the discovery of the planet Neptune.

HSW

3 On what **two** things does the force of gravity depend?

4 Why do moons not fall into the planet they are orbiting?

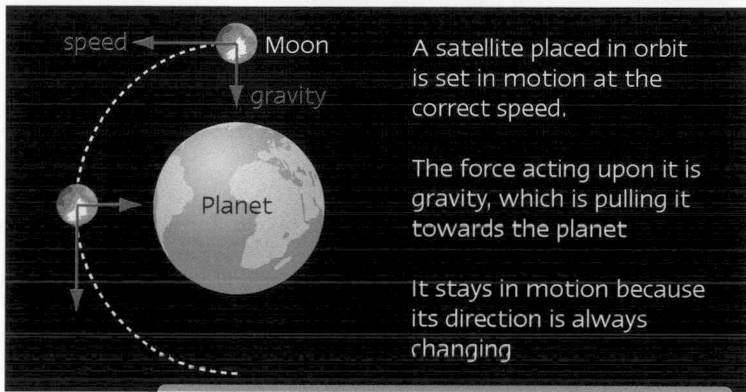

FIGURE 3: How gravity keeps objects in orbit.

Gravity and orbits

Because gravity depends on the mass of the objects, when large objects are involved then the force of gravity can be very strong. Gravity gets weaker as the objects get further apart.

Because gravity depends on the mass of the objects, it plays a very important part in keeping the planets in orbit around the Sun.

Gravity is used to keep **artificial satellites** in orbit. The satellite is positioned at a certain height above the Earth with a forward speed. This enables the satellite to maintain a constant height above the Earth in its gravitational field.

5 On which planet in the Solar System would gravity be strongest?

6 On which planet in the Solar System would the Sun's gravity have the greatest effect?

7 Explain, using your ideas about gravity and orbits, why the planets in the Solar System remain in orbit around the Sun and do not start orbiting each other.

8 Before a satellite can be launched, its orbit needs to be carefully calculated. Why is it important to get these calculations correct?

Did You Know...?

Black holes are formed when massive stars collapse under their own weight. The gravity is so strong that nothing, including light, can escape. In the photo below any surrounding gas and dust are spiralling into the black hole, attracted by the immense gravitational forces. The extreme conditions cause the gas and dust to heat up and glow.

... gravity ... massive ... orbit

History of astronomy

In August 2006, the International Astronomical Union decided that Pluto was not a planet. The evidence suggested that Pluto was not large enough to class as a planet. It is only 1/500th of the size of the Earth.

People's ideas about our Solar System have changed over time. As technology has improved so has our understanding of what the Solar System is.

In ancient times, ideas about the stars were often the foundation of religious beliefs. Different civilisations had their own ideas to explain the apparent movement of the stars across the sky. Stonehenge in Wiltshire is believed to be a very old solar calendar. We think this because of the way the stones and the Sun line up at certain times of the year.

FIGURE 1: Stonehenge.

The Mayans, in South America, developed a very advanced calendar based on observations of the planets, which enabled them to make their human sacrifices to the Gods at the right times of the year.

The ancient Greek model suggested by Ptolemy in about AD 150 put the Earth at the centre of the universe, with some of the planets and the Sun moving around it in circular orbits. This model became the standard model for 1500 years.

FIGURE 2: Ptolemy's model of the Universe.

Throughout the Middle Ages most people believed that the Earth was flat and that you could fall off the edge of it.

A Polish priest called Nicolaus Copernicus (1473–1543) proposed that a rotating Earth revolved around the Sun. This was the first time that a model with the Sun at its centre was proposed.

FIGURE 3: Kepler's model.

Johannes Kepler (1571–1630) adjusted the model so that the orbits of the planets became egg-shaped or elliptical.

Galileo (1564–1642) invented the telescope and his observations with it began to confirm Copernicus' earlier theories.

Isaac Newton (1642–1727), devised a law of gravity which explained the motion of the planets across the sky. During the late 1800s, the field of science desperately needed a new theory to revise the old Newtonian-based physics because it could not explain all the observations that the more advanced telescopes were now making.

Albert Einstein (1879–1955) revolutionised all aspects of science and modern thought through a range of groundbreaking and unique theories which began to explain how gravity, light, time and space are connected.

Stephen Hawking (1942–) has made significant breakthroughs in our understanding of how the universe works by developing a quantum theory of gravity.

Assess Yourself

1. On what evidence did ancient civilisations base their ideas of the Solar System?
2. Where was the Sun in Ptolemy's model?
3. What was the main change made by Copernicus?
4. Why does Kepler's work form the foundation of the modern model of the Solar System?
5. Galileo invented the telescope. Why do you think this made such a difference to our ideas about the planets?
6. Why do you think Einstein's contribution was so important?
7. Name the inner planets and give **three** differences between them and the outer planets.
8. Why does the Moon stay in orbit around the Earth?
9. What causes the seasons?
10. How has our knowledge of the Solar System got better?

ICT Activity

You are going to research some information and produce a timeline about the lives of these and other people who were important in the development of our ideas about the Solar System. You will need to find out when these people published their key ideas, and some more details about those ideas.

Numeracy Activity

Use the Internet to find out the following facts about each planet:

- distance to the Sun (in millions of kilometres)
- average temperature on sunny side (°C)

Draw a line graph to show how the temperature varies with the distance from the Sun.

Which planet does not seem to fit the pattern?

The asteroid belt is about 400 million km from the Sun. Use your graph to estimate the surface temperature of the asteroids.

Level Booster

8 You recognise that improving our understanding of the Solar System and the universe will improve our knowledge about the world around us.

7 Your answers show an advanced understanding of how our ideas about the Solar System have changed and the role of evidence in the development of explanations.

6 Your answers show a good understanding of how our ideas about the Solar System have developed and an appreciation of how the development of these ideas may have impacted on people's lives both now and at the time.

5 Your answers show a good understanding of our planet and the Solar System. You understand about day, night and the seasons. You show a good grasp of some scientific terminology, and some appreciation of how these ideas impacted on people's lives both now and at the time.

4 Your answers show a basic understanding of our ideas about the Solar System and a basic grasp of some scientific terminology.

Gravity and weight

BIG IDEAS

You are learning to:

- Understand the difference between mass and weight
- Explain the relationship between weight and gravity
- Analyse data about gravity and different planets

Mass

The **mass** of an object simply means the amount of material the object is made from. The more material the object is made from, the more mass it has. Things that have a big mass are harder to move and stop than objects with a small mass. So an empty bottle (with only air inside) is easier to move than a bottle filled with liquid. The full bottle contains more material so it has more mass than the empty bottle.

Your mass measures how much material your body has in it. If you eat too much your mass will increase. If you eat less, your mass will decrease. In the metric system the unit of mass is the **kilogram** (kg).

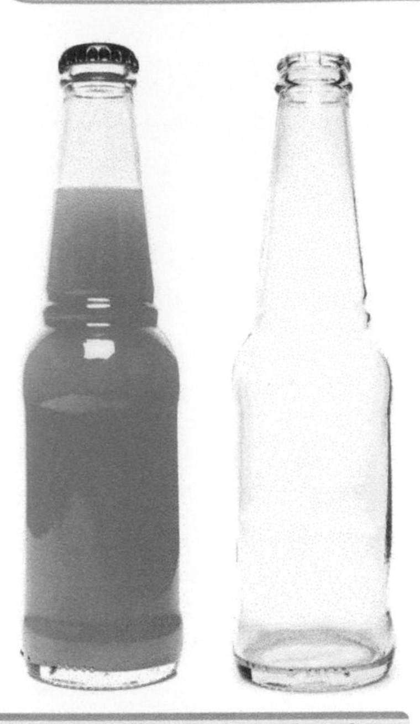

FIGURE 1: Full and empty bottles.

1. What is mass? What units are used to measure it?
2. Why are things with more mass harder to move?

Weight

Because the Earth has a large mass it produces a lot of **gravity**.

Gravity always pulls things down: by 'down' we mean towards the **centre** of the Earth. If this was not the case, people in Australia would fall off the Earth. The hard surface of the Earth stops you from sinking any further towards the centre of the planet. This is just as well.

The **force** of gravity acting on anything gives it **weight**. Weight is a force that is measured in **newtons**.

Gravity on the Earth pulls every kilogram of mass with a force of 10 newtons.

So your weight on the Earth = your mass (in kilograms) × 10

So mass and weight are related.

3. What is weight? What units are used to measure it?
4. What is the weight of a person of mass 65 kg?
5. In which direction does gravity act?

Watch Out!

The confusing thing with weight is that the word is often used incorrectly. For example, we might say that a bag of sugar weighs 1 kg, but this is incorrect. The bag of sugar has a mass of 1 kg but its weight is 10 N.

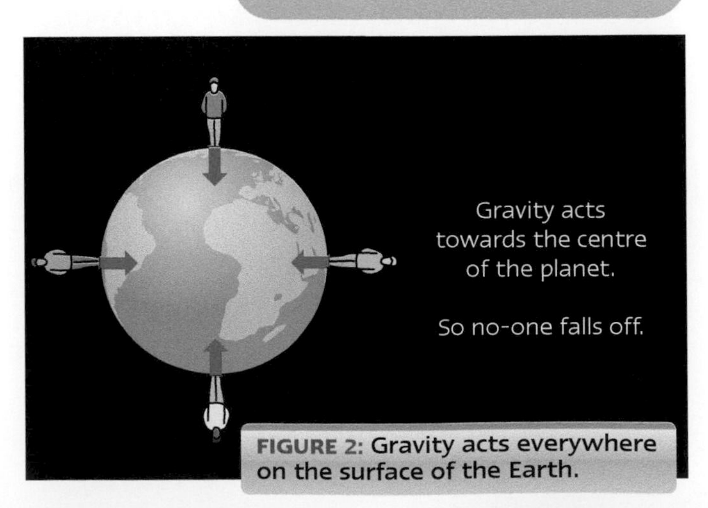

FIGURE 2: Gravity acts everywhere on the surface of the Earth.

... astronaut ... centre ... force ... gravity

Weight on other planets

When **astronauts** travel to the Moon their mass does not change, but their weight will be less.

FIGURE 3: An astronaut jumping on the Moon.

Because the Moon's gravity is about one sixth as strong as gravity on the Earth, you would weigh about one-sixth of your weight on the Earth. You should also be able to jump six times higher on the Moon than you can on the Earth.

Your mass on any planet in the Solar System (Jupiter, Venus, the Earth, or anywhere else) is the same. But if you step on a newton scale to measure your weight, the reading would be different on all of these planets. Your weight will depend on the strength of the force of gravity on each planet.

If you were able to travel to another planet, there are a number of differences you would notice.

	Earth	Moon	Mercury	Venus	Mars
Surface gravity (compared to the Earth	1.0	0.17	0.38	0.90	0.38
Mass (kg)	40	40	40	40	40
How much you can lift (kg)	10	60	30	10	30
How high you can jump (cm)	20	120	53	22	53
How long it takes to fall back to the ground (s)	0.4	2.4	1.1	0.4	1.1

The table above shows the effects of gravity on the Moon and some of the planets in the Solar System.

6 Why does an astronaut weigh less on the Moon than on the Earth?

7 Use the table to work out the weight of a 40 kg person on Mercury and Venus.

8 Draw a line graph of surface gravity against how high you can jump. Is there any pattern in your results? If surface gravity was 0.6 compared to the Earth, how high could you jump?

9 How do scientists know how much you can lift on another planet? How accurate will the figures be?

Did You Know...?

Some scientists think that these organisms closely resemble giraffes, having evolved to feed on extremely tall-growing plants. A rocky planet that is smaller than the Earth would have lower surface gravity. It would also have a thinner atmosphere and would be cooler as a result. If the surface gravity was too small the atmosphere would escape into space. This is the process that many believe occurred on the planet Mars.

FIGURE 4: Low gravity aliens.

Satellites

BIG IDEAS

You are learning to:

- Describe what a satellite is
- Recognise the different types of satellites
- Explain why satellites are useful

Make that call

The word **satellite** is used to describe any object that orbits another larger object in space.

When you make a mobile telephone call to your friends how does the signal reach them?

Most mobile phone calls use satellites to relay the call. Microwaves are used to transmit the call from your phone to a receiver station which then bounces the signal from an orbiting satellite above the Earth to another receiver station which then passes the signal to the receiving phone. It all happens so quickly that you do not notice it. The same thing happens with satellite television. Satellites are used to relay the pictures from one part of the world to another. The signal, which is carried by radio waves, bounces off the satellite and is carefully directed to the receiver.

Sometimes you can see satellites as they move across the sky.

1 What is a satellite?

2 Describe how satellites help you make a mobile phone call.

FIGURE 1: Satellites after and before launch.

What is a satellite?

There are two kinds of satellite, **natural** and **artificial**. The Moon is a natural satellite which orbits a larger object (the Earth), while the Earth, in turn, is a natural satellite of the Sun. Artificial satellites are placed above the Earth's surface and put into orbit above it.

Most artificial satellites are launched from rockets and getting them to the right position above the Earth's surface needs very careful calculations of forces, accelerations and speeds. Artificial satellites have only really become more widely used within the last 30 years. Until then the technology needed was not readily available.

All satellites are held in their orbits by a combination of their speed and the force of gravity.

There are over 300 artificial satellites orbiting the Earth and they have played a major part in the global **communications** revolution.

Did You Know...?

The first satellite placed into orbit was launched by Russia in 1957. It was called Sputnik and it speeded up the development of the Space race which resulted in America reaching the Moon first in 1969.

3 What is the difference between a natural and an artificial satellite?

4 How are most artificial satellites put into orbit?

5 What is meant by the term 'global communications revolution'?

Exam Tip!

You should know some of the different uses of satellites.

Uses of satellites

Satellites have a number of different uses.

- Earth observation – for map making and imaging of geographical features. They are also used in GPS systems. The images are used in the management of the world's environmental resources. GPS satellites are used in car satellite navigation systems.
- Weather forecasting – these enable meteorologists to track weather patterns and to forecast the weather.
- Communications – television, radio and telephone.
- Defence – these include spy satellites, location and guidance satellites.
- Space observation – these are used to observe distant places.
  They are able to provide very clear images.
- Manned satellites and space stations – these research how we can survive in space.

All of these satellites have very sensitive and specialised cameras to enable them to do their job. Many of the instruments that they carry enable them to take different types of photograph.

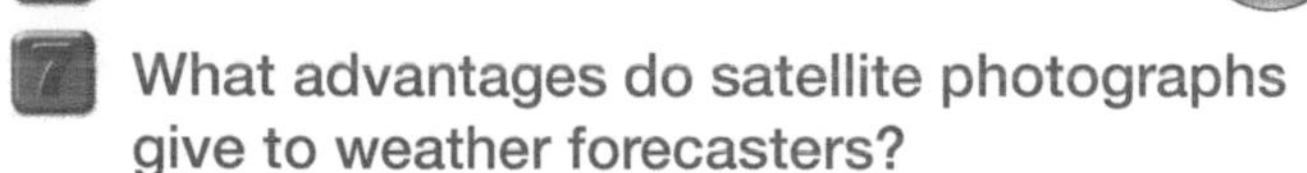

6 Give **three** uses of artificial satellites.

7 What advantages do satellite photographs give to weather forecasters?

How Science Works

In 1990 the Hubble Space Telescope was launched into space. Soon after its launch it was found to be faulty and astronauts had to use the Space Shuttle to repair it.

The Space Shuttle is able to make repeated trips into space but unfortunately there have been several accidents. The Space Shuttle Challenger exploded in January 1986 killing all its crew.

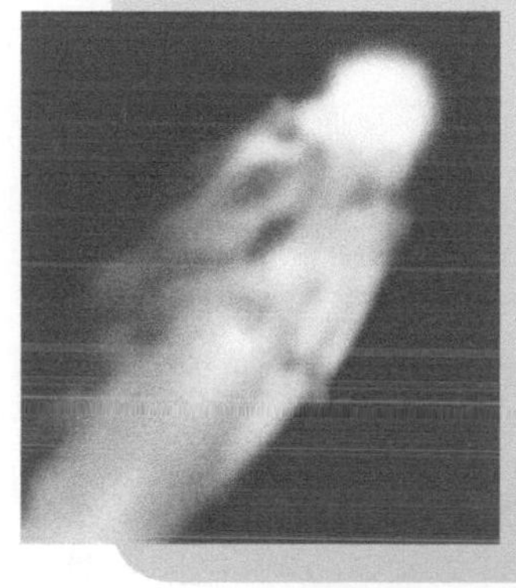

HSW

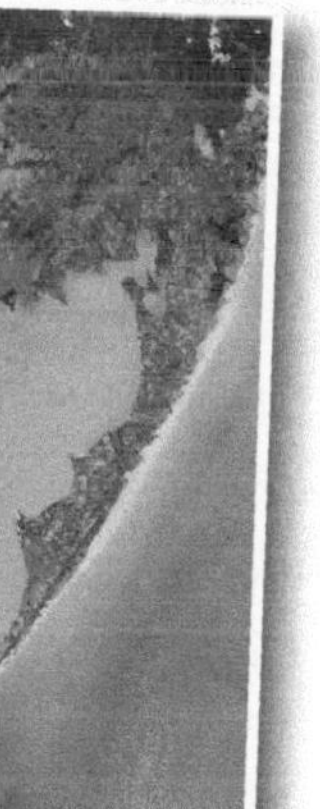

FIGURE 2: Photos of weather patterns taken from satellites could show hurricanes forming.

Staying in orbit

There is only one force acting on a satellite when it is in orbit, and that is the gravitational force exerted on the satellite by the Earth. This force is constantly pulling the satellite towards the centre of the Earth. A satellite does not fall straight back to Earth because its direction is constantly changing.

8 Why do satellites stay in orbit above the Earth?

9 Find out the difference between a geostationary satellite and a polar satellite.

Space travel

BIG IDEAS

You are learning to:
- Understand why people want to travel into space
- Understand how rockets are launched and how they escape the Earth 
- Consider some of the challenges still to be overcome before space travel is a reality for all 

Space, the final frontier

Ever since we first looked at the stars, mankind has dreamt about travelling into space and exploring the **Universe**. What is out there? Where will we go? Is there life?

Space travel has only become a reality in the last 50 years. The first satellite was launched in 1957 and man first set foot on the Moon in 1969. This is still the furthest we have travelled from our planet and no one has returned to the Moon since 1972.

1 Why have people always wanted to travel into space?

The need for space travel

As the Earth's **resources** are placed under growing threat from our expanding **population**, scientists have again begun to consider the possibility of space travel.

2 Why is space travel now being reconsidered?

Launching spacecraft

To launch a spacecraft, it needs to overcome the Earth's gravity. Isaac Newton proposed a thought experiment to show how this could be done. If you throw an object straight up it rises until gravity stops it and it returns to Earth. But, because gravity decreases as you go further from the Earth, if you throw the object so fast that it never quite slows down enough for gravity to pull it back, then it can escape into space.

3 Explain why the shells in Figure 2 eventually fall back to the Earth.

4 What is meant by 'escape velocity'?

Did You Know...?

One of the closest asteroids, Eros, has already been orbited by an unmanned spacecraft. Some asteroids are believed to have massive amounts of natural steel and precious metals. But Type C asteroids contain water.
Water is the key resource. You can drink it, extract oxygen from it, use it to make rocket fuel, and even shield yourself from radiation.

FIGURE 1: Launching a spacecraft.

Imagine firing a very large gun from the top of a very high mountain. The shell would travel a certain distance before falling back to the Earth pulled by gravity, as shown by A.

The shell C is fired at a speed of 17 000 mph. It would still fall towards the Earth, but because the Earth is curved it would stay at the same height above it. The shell would be in orbit. This is the speed that satellites need to be launched at.

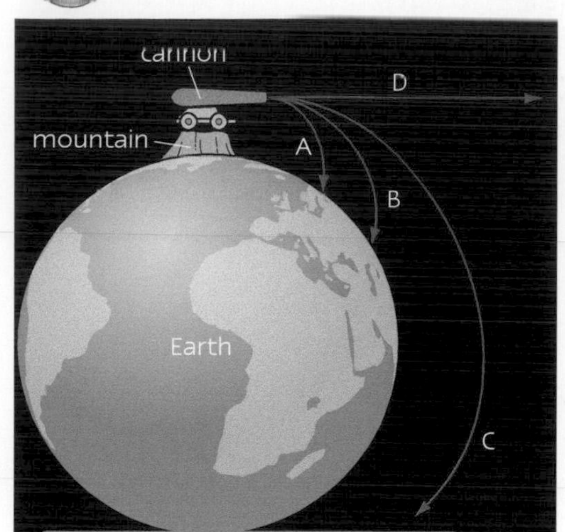

FIGURE 2: Diagram of Newton's thought experiment

B shows the path of a shell fired at a faster speed. It travels further before gravity pulls it back down to the Earth.

If we could fire the shell at a still higher speed of 25 500 mph (7 miles per second!) then it would escape the Earth's gravity and fly off into space. This is shown by D. This speed is called the **escape velocity**. Rockets need a lot of **fuel** to enable them to do this.

To boldly go ...

Launching the rocket into space is only one of many problems that need to be overcome. The other challenges are:

The amount of fuel needed

Any spaceship needs to be able to carry enough fuel to enable it:

- to escape Earth's gravity
- to accelerate to high speed
- to steer during flight
- to land and take off elsewhere
- to keep people alive during the flight.

This much fuel will make the spaceship very heavy. There will need to be some alternative way of providing the energy needed. Solar power and nuclear energy are possible solutions.

The time taken to get anywhere

The maximum possible speed is the speed of light. Scientists do not believe it is possible to travel this quickly. Even at this speed it will take 4.3 years to reach the nearest star (Alpha Centauri) and 100 000 years to reach the nearest galaxy. During this time how will we sustain life?

People need food, oxygen, water, warmth and sanitation. How will these be provided? Humans have a life expectancy of about 70 years so how will we stay alive?

FIGURE 3: This is the largest gun made in the First World War. It was capable of firing a shell 75 miles. The shell was not fired fast enough to escape the Earth's gravity.

How Science Works

A large, lightweight, economical, easy to manufacture human habitation module is being developed.
Transhab is a 3-storey inflatable that can be blown up with the help of nitrogen gas in approximately 10 minutes. At launch, Transhab has a diameter of 14 feet and once it is inflated it will have a diameter of 27 feet and volume of 12 000 cubic feet.

HSW

5 Outline the main challenges that need to be overcome before we can travel into space.

6 Can you think of any other challenges that will need to be overcome before space travel is a reality?

7 Explain the significance of the role of artificial satellites and probes in developing our understanding of the solar system.

Exploring further

BIG IDEAS

You are learning to:

- Understand how telescopes and space probes are used to gather information about distant planets and stars
- Recognise some of the most important space probes ever launched
- Discuss the Hubble Space Telescope and the Cassini-Huygens mission to Saturn

Telescopes

For hundreds of years **telescopes** were the only way in which scientists were able to gather information about the planets and the stars.

We can use telescopes on the ground or in space. In the early days telescopes were not very good at producing clear images.

Since the start of the space age many small spacecraft or **probes** have been sent to some of the planets in the Solar System.

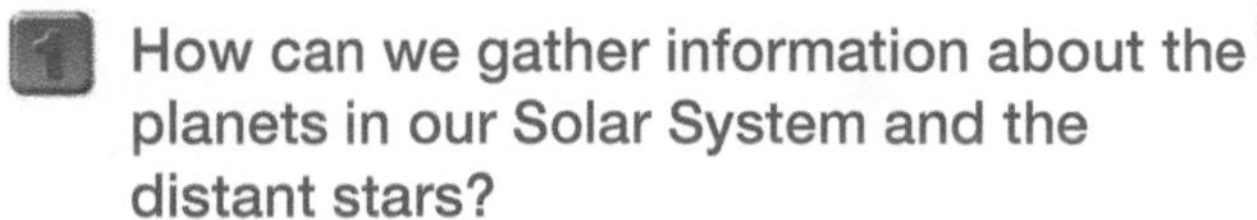

1. How can we gather information about the planets in our Solar System and the distant stars?

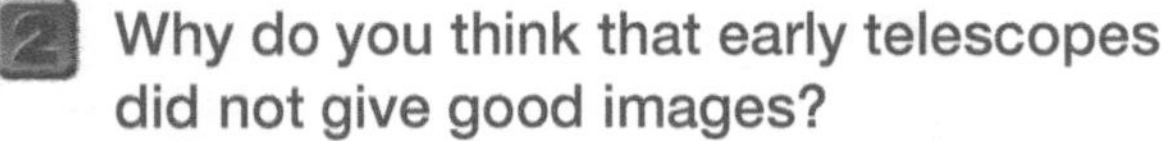

2. Why do you think that early telescopes did not give good images?

FIGURE 1: Space probes.

Unmanned missions

Manned spacecraft have only travelled as far as the Moon but **unmanned** spacecraft or space probes have travelled much further. Because they are unmanned they do not need to carry any of the supplies necessary to sustain life.

Space probes are not designed to return to the Earth. They carry lots of scientific instruments to gather as much information as possible about the places they visit.

Some have landed on other planets! Others have flown past the planets and taken pictures of them. There are even space probes that go into orbit around other planets and study them for a long time.

3. What is a space probe?

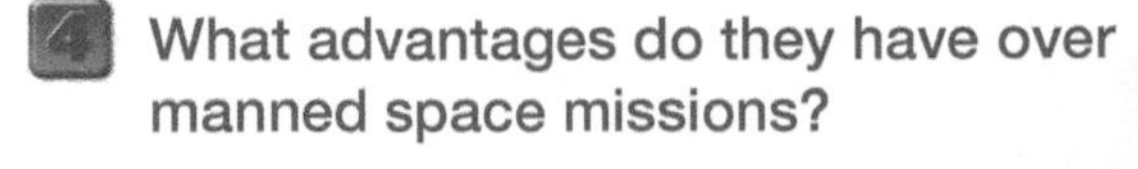

4. What advantages do they have over manned space missions?

Watch Out!

Space probes are designed to gather data and send them back to the Earth for analysis. They are not designed to carry people. Spacecraft such as the Apollo missions are designed to carry people.

... atmosphere ... manned ... probe

Famous space probes

- Pioneer 10. Launched in 1973, it flew past Jupiter and in 1983 it became the first spacecraft to leave our Solar System. It carries a message from mankind. In 2006 we lost contact with it.
- Voyager 1 and Voyager 2 were launched in 1977. They conducted close-up studies of Jupiter, Saturn, Neptune and Uranus.
- Pathfinder was launched in 1996 to study Mars. It used a robot vehicle to survey the planet, take photographs and analyse the rocks and soils on the surface. It took measurements of pressure, temperature and wind speed.

Did You Know...?

Voyager 1 is now the most distant artificial object ever. It is now travelling at the very edge of our Solar System, in an area called the heliosphere. This is where the influence of our Sun becomes virtually zero. Voyager is over a billion miles from the Sun, travelling at a speed of 1 million miles a day!

5 What extra information can a space probe provide that a telescope cannot?

6 Why do you think that Mars is of special interest to scientists?

FIGURE 2: An image from the Hubble Space Telescope.

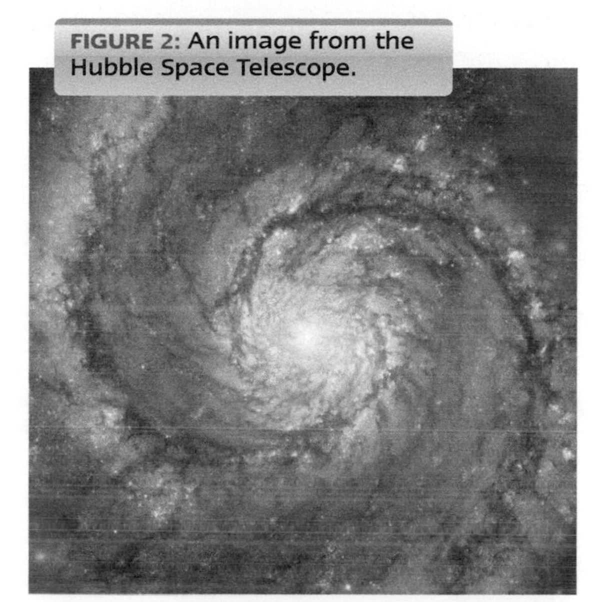

Hubble Space Telescope

The Hubble Space Telescope was launched into the Earth's orbit in 1990. Because it was above the Earth's **atmosphere** it was able to send clear and dramatic photographs of distant objects back to the Earth without the distorting effect of the Earth's atmosphere.

7 What is the advantage of having the telescope in space rather than on the ground?

8 What are the disadvantages?

FIGURE 3: Technicians install the Huygens probe into the Cassini orbiter in July 1997.

Cassini-Huygens

The Cassini probe to Saturn was launched in 1997. It is the biggest and most expensive probe to ever visit another planet. It established orbit around Saturn in July 2004. It will study the planet, its ring system, and many of its moons for at least four years.

The most interesting of Saturn's moons is called Titan. Titan is one of the most mysterious objects in our Solar System. It is the second largest moon and the only one with a thick, methane-rich, nitrogen atmosphere. Experts think that its atmosphere resembles that of the Earth when it was very young. The Cassini craft released a separate probe called Huygens to investigate it.

9 Why is Titan of such interest to scientists?

10 Find out more about the Cassini-Huygens mission and what sort of information it is providing.

Practice questions

1 Complete the following sentences using words from the list. Write the answers in your exercise book.

axis **centre** **complete** **Earth** **Earth** **heat** **orbit**
planets **rotates** **Solar** **Sun**

The third planet out from the Sun is the ________ . The Earth always ________ at the same speed around its ________ . It takes 24 hours to ________ one full rotation. The ________ System is the name given to our ________ and all the ________ which ________ or go round it. The Sun is the ________ of our Solar System and it provides the ________ and light for the ________ .

2 Auckland in New Zealand is on the other side of the world from the UK.

a If it is midnight in London, what time is it in Auckland?

b If it is 6 a.m. in Auckland, what time is it in London?

3 **a** Which is the biggest planet?

b Which **two** planets will be hotter than the Earth?

c How do we see the Moon and the other planets in the Solar System?

d What is an orbit?

4 Choose the best statement.

a You weigh less on the Earth than on the Moon.

b You weigh more on the Earth than on the Moon.

c You weigh the same on the Earth and on the Moon.

5 One advantage of using unmanned spacecraft to explore the Solar System is that they can be easily launched from rockets. Give **two** other advantages.

6 Write these things in order from smallest to largest:

planet **Universe** **Solar System** **galaxy**
asteroid **moon** **star**

7 Complete the following sentences. Use words from the list and write the answers in your exercise book.

axis **seasons** **Southern** **summer**
summer **tilted** **towards** **winter**

We have ________ because the Earth's axis is ________ . In the Northern hemisphere it is ________ when the Earth's ________ is tilted ________ the Sun and ________ when the Earth's axis is tilted away from the Sun. When it is winter in the Northern hemisphere it is ________ in the ________ hemisphere.

8 Artificial space probes are used to explore the Solar System.

a Why have they not explored outside of the Solar System?

b Why have they not carried people?

9 What **two** things does the gravity between any two objects depend on?

10 **a** What does a person of mass 75 kg weigh on the Earth?

b Gravity on the Moon is one sixth of the strength of gravity on the Earth. How much would the same person weigh on the Moon?

c On which planet in the Solar System would the person weigh most? Explain your answer.

d An astronaut hits a golf ball a distance of 900 m on the Moon. How far would it go on the Earth?

11 Satellites can be used for observing space and also for weather forecasting.

a Give **three** other uses of satellites.

b Give **two** of the benefits of using satellites for weather forecasting.

c Explain why a satellite stays in orbit above the Earth.

12 Choose the correct statement.

a Satellites stay in orbit because there is no air resistance in space.

b Satellites stay in orbit because there is no gravity at that height above the surface.

c Satellites stay in orbit because they are moving at high speed in a straight line.

d None of the above.

13 Why have scientific ideas about the Solar System changed over time?

14 Explain, giving examples, the significance of the role of artificial satellites and probes in developing our understanding of our solar system?

Learning Checklist

- ☆ I know what a telescope is used for. page 204

- ☆ I know why we have day and night. page 186
- ☆ I can name the planets in the Solar System. page 192
- ☆ I know that gravity is the force which acts on objects and pulls them towards the centre of the Earth. page 194
- ☆ I know the force of gravity on objects is called weight. page 198
- ☆ I know that I would weigh less on the Moon than I do on the Earth. page 199
- ☆ I can describe why unmanned space probes are used. page 204

- ☆ I can describe why we have the seasons. page 188
- ☆ I can describe the phases of the Moon. page 190
- ☆ I know that gravity acts everywhere in the Solar System including on the Earth or on the Moon. page 194
- ☆ I know that mass and weight are different and I can give examples of each. page 198
- ☆ I know the difference between a natural and an artificial satellite. page 200
- ☆ I know why we are exploring space. pages 202, 204

- ☆ I understand that gravity pulls all objects together and what the size of the force depends on. page 194
- ☆ I can describe how mass and weight are different and how gravity causes weight. page 198
- ☆ I can use information to compare gravity and explain why weight is different on different planets. page 199
- ☆ I can describe some uses of satellites. page 200
- ☆ I can describe how spacecraft are launched. page 202
- ☆ I know some of the benefits of using a space telescope. page 204

- ☆ I understand the relationtionship between evidence and scientific ideas and recognise that these are applied to develop our understanding of the structure of the Solar System. page 192
- ☆ I can explain how gravity keeps planets and satellites in position. page 194
- ☆ I can describe how the forces on rockets or satellites change as they travel away from the Earth. page 202
- ☆ I can explain some of the factors that limit space exploration. pages 202, 204
- ☆ I understand why the relationship between ideas and evidence changes and can apply this to the role of artificial satellites and probes in space exploration. page 205

Topic Quiz

1 What are telescopes used for?

2 Why do we have day and night?

3 Name the planets in the Solar System outwards from the Sun.

4 Give **two** reasons why unmanned space probes are used to explore space.

5 Why are the days longer and warmer in summer than in winter?

6 What is the difference between a natural and an artificial satellite?

7 What is the difference between mass and weight?

8 What is the difference between a solar and a lunar eclipse?

9 Give **three** uses of satellites.

10 Give **two** benefits of space telescopes.

11 Why does a satellite stay in orbit?

12 Give **three** factors that limit how far we can travel in space.

True or False?

If a statement is false then rewrite it so it is correct.

1 The Earth rotates around its axis every 12 hours.

2 Telescopes are used for studying distant planets and stars.

3 There are seven planets in the Solar System.

4 We have seasons because the Earth is tilted on its axis.

5 Gravity stays at the same strength everywhere in the Solar System.

6 Mass is caused by the force of gravity pulling things towards the centre of the planet.

7 Satellites are not used for radio communications.

8 In a lunar eclipse the Moon moves into the shadow of the Earth.

9 The distance we can travel in space is only limited by the speed we can reach.

ICT Activity

You are going to undertake some research into the planets in the Solar System. Choose any **four** of the planets within the Solar System. For each one find out at least **five** different facts about the planet and include a photograph of it. Which of your planets have been visited by a probe which has landed and when; and which ones have not? For those that have been landed on, try to find an image taken by the probe which explored it. For those that have not been visited, try to find out if there are any plans to visit them in the future.

Produce a leaflet, poster or PowerPoint presentation to show what you have found out.

Alternatively, you could produce a timeline which shows the main events in space exploration since 1957. Illustrate it with photographs and key facts.

Geography Activity

Satellite images are very useful in geography. Consider what sort of images and information is useful for geographers to know and how satellites are providing this information. You may choose to look at the weather, farming patterns, geology, urbanisation, pollution or any other suitable area. Produce a report to say what you have found out.

Glossary

Keyword	Definition	Page
Acid rain	Rain that has a pH of less than 7	120
Alchemist	A medieval chemist who attempted to turn metals like iron into gold	68
Allergy	An irritating immune response to things that are not harmful, like pollen	34
Alloy	A mixture of metals	60
Amplitude	Half the distance between the trough and the crest of a wave	134
Antibodies	The molecules made by the immune system that recognise microbes and help get rid of them	36
Antitoxins	Molecules made by the immune system that neutralise toxins	36
Atmosphere	The mixture of gases above the Earth	120, 204
Atomic number	Number that shows how many protons an atom has	84
Biological control	Using a predator to control a pest	178
Boiling point	The temperature at which a liquid turns into a gas	72
Brain	The organ that controls the rest of the body	144
Brittle	A material like graphite that snaps when it breaks because it is not stretchy or bendy	84
Carbon dioxide	A gas green plants use to make food and animals breathe out. It is made by reacting carbon with oxygen	14
Chemical reaction	A reaction between two or more elements or compounds which causes chemical change	92
Chemical symbol	The abbreviation used to represent a chemical element	82
Chromatogram	A collection of substances separated by chromatography	74
Chromatography	A method used to separate soluble substances	75
Compost	Decayed plant material	174
Compound	A substance made when two or more elements are chemically joined	90
Compression	Part of a wave where particles are closer together	138
Conductor	Material that lets heat or electricity pass through it	154
Corrode	What acids do when they react with metals and make them dissolve	68

Current	What passes through a material that conducts electricity	110, 112
Decay	Break down	174
Decibel (dB)	The unit of loudness	146
Decomposers	Organisms that break down plant or animal tissue	174
Dilution	Mixing with water	48
Disinfectants	Solutions that destroys microbes	49
Dissipate	Spread into the surroundings	136, 138
Distillation	Separation of a liquid from a mixture by evaporation and condensation	66
Eclipse	When the light from the sun is blocked	190
Element	Pure substance with only one type of atom	82
Energy	Something that can cause change, like heat or light	138, 147, 150
Force	A push, pull or twist	198
Formula	A code that uses little numbers to show the ratio of elements in a compound e.g. CO_2 shows that carbon dioxide contains 2 oxygen atoms for every carbon atom	92
Frequency (f)	Number of vibrations or waves per second	134, 142
Fuel	A substance that releases energy when it burns	202
Gas	A material made of separate moving particles	140, 154
Global warming	The gradual temperature rise in the Earth's atmosphere and oceans	121
Gravity	The force that pulls things towards each other	186, 190, 192, 195, 198
Habitat	The place where something lives	176
Hearing range	The range of frequencies an animal can detect	142
Hemisphere	Half of a sphere like the Earth	188
Hertz (Hz)	The unit of frequency	134
Immune system	The parts of our bodies that resist disease	36
Incubation period	The time before the symptoms of a disease appear	34
Inoculate	Inject someone with weakened microbes in order to induce immunity to a disease	44
Insoluble	Does not dissolve	62, 65
Insulator	A material that does not transmit heat or electricity	154
Kinetic energy	The energy something has because it is moving	115
Liquid	A material that flows	140, 155
Lunar eclipse	When the Earth blocks light from the Sun and makes a shadow on the Moon	190

Malleable	Can be bent into different shapes	69, 85
Mass	The amount of matter in something	199
Microscope	A device used to make magnified images	83
Mixture	Two or more substances that are not chemically joined	89, 91
Newton	The unit of force	199
Non-ferrous	Not made of iron	83
Nutrients	Chemicals needed in a balanced diet	175
Oscilloscope	A device that can convert a sound into a wave image	134
Oxide	A compound made by reacting an element with oxygen	91, 93, 94
Oxygen	A gas produced by plants and used by animals and burning materials	14, 93
Particle	A small piece of material e.g. an atom or a molecule	138, 140, 155, 157
Pathogens	Organisms which cause disease	35
Periodic Table	A table that lists every element	85
Pesticide	A chemical used to kill pests	179
Pitch	The frequency of a sound	135, 142
Pollution	Damage caused by pollutants	121
Population	The number of individuals present	176, 202
Predator	An animal that catches and kills other animals for food	177
Prey	An animal that is hunted and killed by other animals	177
Product	Something produced by a reaction	15
Push wave	A wave-like sound made by pushing on something	139
Rarefaction	Part of a wave where particles are further apart	139
Ratio	The link between two values e.g. if the first value is twice as big the ratio is 2:1	95
Reflect	Bounce off something	137, 155, 191
Respiration	The release of energy from glucose and oxygen in living things	15, 19, 175
Salinity	The amount of salt in water	73
Satellite	A device that orbits a planet	191, 201
Season	Summer, autumn, winter or spring	188
Signal generator	A device that produces a signal	135
Smog	Fog mixed with pollutant gases	121
Solar eclipse	When the light from the Sun is blocked by the Moon	191
Solid	Made of particles that cannot move around	140, 155

Term	Definition	Page
Soluble	Able to dissolve	60, 63
Solvent	A liquid that something will dissolve in	63
Solvent front	The furthest distance a solvent has moved during chromatography	75
Sonar	Uses sound reflections to detect things underwater	137
Sucrose	Common sugar	61
Symptoms	Signs that someone has a disease	35
Telescope	A device that uses lenses to magnify distant objects	193, 205
Toxins	Things that damage a living organism	35, 37
Transmit	Send	139
Vaccine	Something that provides immunity from a disease	45
Vacuum	A place with no particles in it	141
Vapour	Gas evaporating from a liquid	69
Wave	How light or sound energy is transmitted	123, 139
Wavelength	The distance between two crests or troughs on a wave	135
Weight	The force of gravity on an object	199
White blood cells	The cells in the blood that fight disease	37

Acknowledgements

The Publishers gratefully acknowledge the following for permission to reproduce copyright material. Whilst every effort has been made to trace the copyright holders, in cases where this has been unsuccessful or if any have inadvertently been overlooked, the Publishers will be pleased to make the necessary arrangements at the first opportunity.

Cover photograph from NASA / Science Photo Library

The Publishers would like to thank the following for permission to reproduce photographs:

p. 6 © David Munns / Science Photo Library, © Dmitry Nikolaev / istockphoto.com; p. 8 © Michael Cole / CORBIS; p. 9 © Hannamariah photography / istockphoto.com; p. 10 © Christine Balderas, © Andres Peiro Palmer, © 20 08 Jupiterimages Corporation, © Andrew Manley, © Dan Chippendale, © Maria Gritcai, © Malgorzata Korpas, © KMITU / istockphoto.com; p. 11 © Krzysztof Kwiatkowski, © Thomas Perkins; p. 14 © Eliza Snow, © Barbara Tarabura; p. 15 © Jaroslaw Wojcik, © pamspix / istockphoto.com, © Michal Rozanski, © 2008 Jupiterimages Corporation, © Michael Kemter, © Jacom Stephens, © Chris Schmidt, © Andrew Howe, © Adrian Albritton; p. 16 © John Shepherd / istockphoto.com, © Kativ / istockphoto.com; p. 21 © CNRI / Science Photo Library; p. 22 © Roman Milert, © Timothy Wood; p. 23 © Faye Norman / Science Photo Library; p. 24 © 2008 Jupiterimages Corporation; p. 25 © moodboard / Corbis; p. 29 © Brasil2 / istockphoto.com; p. 30 © Michael Pettigrew, © Andy Crump, TDR, WHO / Science Photo Library, © Alex Gumerov, © Biophoto Associates / Science Photo Library; p. 31 © Vanessa Vick / Science Photo Library; p. 32 © Rainer Hillebrand, © Science Photo Library; p. 33 © Science Photo Library, © Science Photo Library, © Science Photo Library, © Biomedical Imaging Unit, Southampton General Hospital / Science Photo Library; p. 34 © 2003 Getty Images; p. 35 © Sebastian Kaulitzki, © Hasan Kursad Ergan, © Martin Firus, © Maciej Laska; p. 36 © Suzanne Tucker, © sgame / istockphoto.com; p. 38 © arlindo71 / istockphoto.com, © Dr M.A. Ansary / Science Photo Library, © CNRI / Science Photo Library; p. 40 © Monika Wisniewska, © David Scharf / Science Photo Library, © Roger Harris / Science Photo Library, © Stephen Sweet, © Science Source / Science Photo Library, © Carolina K. Smith, M.D., © Eliza Snow; p. 42 © Graeme Purdy; p. 43 © Loren Rodgers; p. 44 © CDC / Science Photo Library, © Sheila Terry / Science Photo Library; p. 45 © National Library of Medicine / Science Photo Library; p. 46 © Andreas Reh, © dra_schwartz / istockphoto.com; p. 47 © Sean Warre; p. 48 © alohaspirit / istockphoto.com, © Gary Martin; p. 49 © dra_schwartz / istockphoto.com; p. 50 © Luis Carlos Torres, © Mikael Damkier, © Marc Dietrich; p. 53 © Jean Frooms; p. 56 © Silviu Hisom. Image from BigStockPhoto.com, © les polders / Alamy, © Nicu Mircea. Image from BigStockPhoto.com; p. 57 © Florida Images / Alamy; p. 58 © blickwinkel / Alamy; p. 59 courtesy of www.mothershiptonscave.com; p. 60 © Alan Dunlop-Walters. Image from BigStockPhoto.com, © Ewa Brozek. Image from BigStockPhoto.com, © Elena Schweitzer, © Piotrek Golebiowski; p. 61 © Graça Victoria. Image from BigStockPhoto.com; p. 62 © Henri de Foucauld, © Mike Cryer; p. 63 © Charles D. Winters / Science Photo Library; p. 64 © Deborah Benbrook. Image from BigStockPhoto.com, © Darrin Jenkins / Alamy, © Ferran Traite Soler; p. 65 © Mary-Ella Keith / Alamy; p. 66 © Benoit Beauregard, © James Steidl; p. 69 © sciencephotos / Alamy; p. 70 © Pablo Paul / Alamy; p. 72 © Helene Rogers / Alamy; p. 73 © Alexander Hafemann, © Giovanni Rinaldi; p. 80 © Adrian Beesley, © Ian Wilson, © Vadim Ponomarenko; p. 81 © Dragan Trifunovic; p. 82 © Armando Frazao; p. 83 © Peter Spiro, © Milos Luzanin. Image from BigStockPhoto.com, © Drs A. Yazdani & D.j. Hornbaker / Science Photo Library, © Drs A. Yazdani & D.j. Hornbaker / Science Photo Library, © Drs A. Yazdani & D.j. Hornbaker / Science Photo Library, © Drs A. Yazdani & D.j. Hornbaker / Science Photo Library; p. 84 © Timothy Large, © Mike Hill / Oxford Scientific; p. 85 © Mehau Kulyk / Science Photo Library, © Andrew Lambert Photography / Science Photo Library; p. 88 © Elnur Amikishiyev. Image from BigStockPhoto.com, © Thomas Mounsey. Image from BigStockPhoto.com; p. 89 © Olga Khoroshunova. Image from BigStockPhoto.com, © Richard Backlund, © AA World Travel Library / Alamy; p. 90 © Don Wilkie, © Raul Gonzalez Perez / Science Photo Library; p. 91 © Andy Clarke / Science Photo Library, © Bernhard Edmaier / Science Photo Library, © Bernhard Edmaier / Science Photo Library; p. 92 © Gaja Snover. Image from BigStockPhoto.com, © Charles D. Winters / Science Photo Library; p. 93 © Izzat Bakhadyrov. Image from BigStockPhoto.com; p. 95 © ALIKI SAPOUNTZI / aliki image library / Alamy, © kim collison – Fotolia.com, © Guy Erwood. Image from BigStockPhoto.com; p. 100 © Victor Habbick Visions. Image from BigStockPhoto.com; p. 101 © 2008 Jupiterimages Corporation; p. 103 © Glen Teitell, © Mark Evans, © Marcus Clackson, © Matt Jeacock, © Paul Ridsdale / Alamy; p. 104 © Cordelia Molloy / Science Photo Library; p. 107 © Roman Krochuk, © Sally Bensusen / Science Photo Library; p. 108 © Martyn F. Chillmaid / Science Photo Library; p. 110 © Alex Slobodkin, © rgbdigital.co.uk – Fotolia.com; p. 112 © Ugur Bariskan; p. 114 © Chris Elwell, © 2008 Jupiterimages Corporation, © 2008 Jupiterimages Corporation, © Eugene Tochilin. Image from BigStockPhoto.com; p. 116 © China Images / Alamy; p. 118 © Paul Gibbings. Image from BigStockPhoto.com, © 2008 Jupiterimages Corporation; p. 119 © ekspansio.com / istockphoto.com; p. 120 © 2008 Jupiterimages Corporation; p. 121 © Joe Pasieka / Science Photo Library, © Survival Anglia / Oxford Scientific, © Mike Hill / Oxford Scientific; p. 122 © 2008 Jupiterimages Corporation; p. 123 © Martin Bond / Science Photo Library; p. 124 © Eva Serrabassa, © US Department Of Energy / Science Photo Library, © 2008 Jupiterimages Corporation; p. 125 © US Department Of Energy / Science Photo Library; p. 130 © René Mansi, © US Department Of Defense / Science Photo Library; p. 131 © Jeremy Walker / Science Photo Library, © Jim Parkin; p. 132 © 2008 Jupiterimages Corporation; p. 133 © 2008 Jupiterimages Corporation, © 2008 Jupiterimages Corporation; p. 134 © Chen Chun Wu, © Index Stock / Alamy, © Andrew Lambert Photography / Science Photo Library; p. 136 © ETCbild / Alamy, © Soubrette / istockphoto.com; p. 139 © 2005 Masa Ushioda / Image Quest Marine; p. 140 © Simon Podgorsek; p. 140 courtesy of the author; p. 140 © NASA / Science Photo Library; p. 141 © Pasquale Sorrentino / Science Photo Library; p. 142 © Keri Lummis – the Kennel Club; p. 143 © Ethan Janson DDS / www.seattle-dentist.com, © Ethan Janson DDS / www.seattle-dentist.com, © Gustoimages / Science Photo Library, © Paul Rapson / Science Photo Library, © Victor Habbick Visions / Science Photo Library; p. 144 © BSIP VEM / Science Photo Library, © M. Dauenheimer, Custom Medical Stock Photo / Science Photo Library, © Bo Veisland / Science Photo Library; p. 145 © Steve Gschmeissner / Science Photo Library, © CNRI / Science Photo Library, © Anatomical Travelogue / Science Photo Library, © Gunilla Elam / Science Photo Library, © Steve Gschmeissner / Science Photo Library; p. 146 © Susan Stevenson, © Jane Shemilt / Science Photo Library; p. 148 © Manuela Weschke; p. 152 © Jon Arnold Images Ltd / Alamy, © imagebroker / Alamy, © Pascal Goetgheluck / Science Photo Library; p. 153 © Matjaz Boncina; p. 154 © Aaron Flaum / Alamy, © Alina Goncharova. Image from BigStockPhoto.com; p. 155 © Ken Babione; p. 156 © Bubutim / istockphoto.com, © Elemental Imaging. Image from BigStockPhoto.com, © Andreas Resch. Image from BigStockPhoto.com, © Thomas Hansson. Image from BigStockPhoto.com; p. 158 © Daniel Cooper; p. 159 © Wiktor Bubniak, © SHAMIL ZHUMATOV / Reuters / Corbis, © Alfred Pasieka / Science Photo Library; p. 164 © Andrew Howe, © Pawel Gaul; p. 165 © Bronwen Sexton, © Science Photo Library; p. 167 © Cat London, © Karen Givens, © arlindo71 / istockphoto.com, © Pavel Lebedinsky, © Steve Mcsweeny; p. 168 © Liz Leyden, © Chris Fourie; p. 169 © Niels van Gijn; p. 170 © Tomasz Zachariasz, © Chartchai Meesangnin; p. 172 © Hansjoerg Richter, © Josh Webb; p. 174 © Tom England, © Jim Jurica; p. 175 © Foto Pfluegl / istockphoto.com, © Paul Hebditch; p. 176 © Nick Tzolov; p. 177 © Tammy Peluso, © Roman Krochuk, © Tatiana Morozova; p. 178 © Dieter Hawlan; p. 179 © Ana Vasileva, © Eric Delmar; p. 180 © Liz Leyden; p. 183 © Thomas Lindner; p. 184 © Digital Art / CORBIS, © Magrath / Folsom / Science Photo Library; p. 185 © David Hardy / Science Photo Library, © Peter Menzel / Science Photo Library; p. 186 courtesy of NASA Headquarters – GReatest Images of NASA (NASA-HQ-GRIN); p. 187 © Planetary Visions Ltd / Science Photo Library; p. 188 © Karl Dolenc, © Christopher Steer; p. 190 © Dennis Hallinan / Alamy; p. 191 © Dr Fred Espenak / Science Photo Library; p. 192 © Friedrich Saurer / Science Photo Library; p. 193 © US Geological Survey / Science Photo Library, © Detlev Van Ravenswaay / Science Photo Library; p. 194 © Leona Barratt, © Joanne Green; p. 195 © Christian Darkin / Science Photo Library; p. 196 © John Evans. Image from BigStockPhoto.com, © 2008 Jupiterimages Corporation, © Science Photo Library; p. 198 © jallfree / istockphoto.com; p. 199 © NASA Images / Alamy, © Christian Darkin / Science Photo Library; p. 200 courtesy of NASA Johnson Space Center (NASA-JSC), courtesy of NASA Johnson Space Center (NASA-JSC); p. 201 © NASA / Science Photo Library, © NASA / Science Photo Library, courtesy of NASA Jet Propulsion Laboratory (NASA-JPL), © Vladislav Gurfinkel. Image from BigStockPhoto.com; p. 202 courtesy of NASA Marshall Space Flight Center (NASA-MSFC); p. 203 © Hulton-Deutsch Collection / CORBIS; p. 204 © NASA / Science Photo Library, © Stocktrek / Corbis, © Jim Sugar / CORBIS; p. 205 courtesy of NASA Goddard Space Flight Center (NASA-GSFC), courtesy of NASA Kennedy Space Center (NASA-KSC)